MOVEMENT
for the
Performing Artist

James Penrod
University of California, Irvine

Illustrations by ROBERT CARR

 Mayfield Publishing Company

Dedicated to Virginia Zobell Heileson

Library of Congress Catalog Card
Number: 73-93344

International Standard Book Numbers:
    0-87484-234-4 (paper)
    0-87484-235-2 (cloth)

Manufactured in the United States
of America

Mayfield Publishing Company
285 Hamilton Avenue
Palo Alto, California 94301

The author wishes to acknowledge the
help of Adrienne Fisk in the preparation of the manuscript.

This book was set in Caledonia by
Typographic Service Company.

The editors were C. Lansing Hays and
London Green; Michelle Hogan super-
vised production; designer of the text
and cover was Nancy Sears.

Contents

iii

*Love the art in yourself, rather
than yourself in art.*

CONSTANTIN STANISLAVSKI

Introduction

This book is written to share my own experiences as a teacher, choreographer, and performer working with actors, dancers, and singers. Some of the ideas are my own, while others I have learned from my own teachers, my students, and other performers and writers.

It is written for performers of all kinds: actors, dancers, and singers. Many of the movement problems are the same for all three, and indeed in the modern theatre world many performers are required to fulfill all three functions.

The book attempts to help you do four things: to look at movement not with the eyes only but with *total* perception; to discover the psychological and physical sources of movement in yourself and in others; to attain intellectual and physical control over your movements; and finally to apply these skills to the development of a role.

v

Throughout this book I have used the literary convention of "he" for both men and women. I regret the inconvenience it may cause women readers.

Neither books nor other people can teach you to be an artist or guarantee that you will become proficient in the practice of your craft. Books, teachers, directors, and fellow performers can only give guidance and insights into working methods that have been successful for others and might be so for you. Teachers and directors can guide you through specialized techniques related to an art form, such as exercises for voice development or the dance. They may also be able to assist you in developing expressiveness in the practice of your craft. However, the ultimate responsibility for artistic achievement lies with you, the individual. You are the only person who can, through hard work, study, observation, intelligence, and talent, develop your technique. And you are the only person who can transcend your technique to express both your own feelings and a view of the human condition.

As human beings we communicate with one another in several kinds of language. One of the major languages is, of course, speech. Equally important, however, is the *silent* language: the language of body movement. Consciously or unconsciously we respond to the physical messages sent by others: almost imperceptible movements of the face, subtle shifts of the body, and many expressive gestures of other kinds. The performing artist, whether actor, singer or dancer, likewise uses the silent language to communicate with his audience. Therefore learning to see movement and to use movement expressively is an important part of the craft of performance. Whether your purpose is to become a better performer or only a more perceptive human being, you will find that learning both to read the silent body language of others and to speak your own silent body language is an exciting and rewarding undertaking.

Often beginning performers panic when they are asked to move in a formal manner: "Start on your upstage foot," or "Keep your body in alignment so that you can be seen and heard." Acute self-consciousness can and frequently does develop: dangling at the ends of the arms are suddenly heavy and extremely awkward hands.

Unfortunately, much of our cultural and educational thrust has been to repress self-expression through movement. Children, who move spontaneously, are often required to sit still, to stop wiggling, to control themselves. A tight control of movement patterns is too often considered "civilized," while large, exhuberant gestures are sometimes considered proper only to the very young, the uncouth, or the primitive. Training in movement is too often limited to sports activities and grammar school folk dance. With so little formal training in body movement, it is no wonder the beginning performer feels awkward. Indeed, he may be embarrassed to pay such attention to that thing known as his body, which he has blindly occupied for so many years. Nonetheless, that is precisely what a performer must do. A body is one of the tools of the trade. How do you overcome a lifetime of learning to ignore and inhibit the very body that is so vital in communication? You must learn to see movement again, in yourself and in others. You must awaken your senses and be keenly aware of them. You must explore your emotions as sources of physical movement. You must analyze movement for its mechanical, anatomical, spatial, and dynamic content, and then bring it under conscious control. You have to re-sensitize, re-awaken, "become alive," and know it! You must train and use your body to its fullest expressive potential.

It is the intent of this book to help you overcome inhibitions you may have about moving, discover the sources of human movement, and learn to organize movement so that you become a more expressive performer. For these reasons you are encouraged to try all the exercises that follow to see if they work for you.

1 Experiencing Movement

As human beings we move constantly and see others moving, but few of us ever really analyze or even take note of the remarkable sensations of motion. However, as a performer you can become both more expert and more expressive by consciously noting the physical sensations of your own movement and analyzing the effect of your own and others' motions.

BODY ASSESSMENT

At first, to develop physical awareness of yourself, perform a general assessment of your own body. What does it look like? How is it put together? How does it move in space? The general shape of your body is largely determined by your skeletal framework: the head, the whole

1

torso (which includes the shoulders, rib cage, and pelvis), and the legs and arms. The size of each part relative to the whole determines whether you are tall or short, barrel-chested, narrow-waisted, and so forth. Attached to the skeleton are the muscles and tendons that provide the force to move and to stabilize the skeleton. They also shape the contours and planes of the body. Stretched over all of this is the skin, which of course differs in tone and shade from person to person. Nothing practical can be done to reshape your skeleton, but the muscles and tendons can be stretched and strengthened to enhance their aesthetic appeal, and of course intelligent dieting can also alter the weight and shape of your body.

In developing body awareness, it is helpful to think of the body architectonically. The head is perched on the long, flexible, segmented column of the spine. At the lower end of the spine is the round or heart-shaped pelvis, to which the bones of the upper leg are attached. The lower leg bones lead to the heel bones and to the arch and toe bones. All the weight pressing down from above is borne on the tiny supports of the feet. Below the neck are the yoke-shaped bones of the shoulders, from which hang arm and hand bones. The skeletal framework of the body is constructed on the principle of a machine with levers that can be raised, lowered, bent, straightened, or rotated. The bone is a lever, the joint is a fulcrum, and the muscles apply the force to move them. In order to give the skeleton stability and the power to move, muscles and tendons are attached at strategic points. At the command center of this remarkable living machine is the brain. Protected within the rib cage or pelvis are the vital organs that serve or maintain the body functions.

After you have studied your own body, look at other people and try to describe how their bodies are shaped: a large head on a small body, a short waist, long arms, short legs, and so forth. None of what might be called "shortcomings" need be a hinderance in the theatre. Anyone who wants to can disguise or take advantage of any so-called disproportion through diet and proper body carriage and movement.

Beyond that, each of us is basically confined to the body he was born with. Diversity of types is the rule in life and, as theatre reflects life, diversity is desirable there, too. With body carriage and acting techniques, makeup, and lighting, one can achieve in the theatre almost any image desired.

EXPERIENCING MOVEMENT KINESTHETICALLY

We all possess three senses through which we are aware of our movements and body. They are the kinesthetic sense, the static sense, and the visceral sense. The kinesthetic sense is important in determining and controlling both your body positions and your movement. It helps you to stand erect, walk, talk, and perform other motor skills. The static sense gives you information about the position of your body in space, the direction of your movement in space, and the changes in speed of a movement. The visceral sense gives you information about the functioning of the internal organs. It cues you when you are hungry, thirsty, or full, and on such matters as the state of the bladder and bodily tension. Hereafter, for our purposes all three senses will be referred to comprehensively as the kinesthetic sense. Your kinesthetic ability makes it possible to do a whole range of activities, from threading a needle to sinking a basketball through the hoop to lifting a pencil or a newborn baby. Other kinds of kinesthetic responses are the physical sensation you experience when you see someone take a bad fall or when you go up or down in an elevator.

The discussion which follows is intended to help you to experience motion kinesthetically: to be aware of the sensation of movement, where and how the body is placed in space, and how much energy is being used during a given movement. To be kinesthetically aware is, in short, to be aware of what every part of your body is doing. The development of the kinesthetic sense is important to any performer in several ways. For example, stage movements are too often initiated externally rather than motivated from within, so that they never take on a realistic vitality and are often incomplete, stiff, and erratic, rather than broad, free, and convincing. This often causes a conscious or unconscious dissatisfaction in the audience. If, however, you can develop a kinesthetic sensitivity, the majority of your movements will take on conviction. You will also be able to learn patterns of movement in much less time.

A number of exercises that are sometimes used in therapy or dance to develop the kinesthetic sense follow. Although they seem relatively simple they do require a high degree of concentration, relaxation, and sensitivity to your own body, so that you can experience yourself as an entity of a certain shape and size occupying space and functioning as a living organism.

EXERCISE 1

Begin by lying on your back with one hand placed on your diaphragm just below the ribs and the other hand placed on your chest. Breathe in and out naturally. Be aware of the rise and fall in your diaphragm. Try to sense the air as it enters your body and fills the lungs and then is expelled. Next breathe in and hold the breath for a slow count of five

3

and then breathe out. Repeat the same action several times, becoming aware of the kind of effort and tension required to breathe in and hold the breath, and of the sense of relief and relaxation when breathing out.

EXERCISE 2

Next, still lying down, close your eyes and imagine the space within your own body as being hollow; then concentrate on the relationship between the shape of your body and the space around you. Take a mental inventory of where all of the parts of your body are in relationship to one another and to the space they occupy. Be aware of how the various parts of your body feel in contact with the floor or the space around you. Be aware of tensions, of your breathing, and of any other sensations or activities taking place in your body. Next, try to suspend conscious control over your body, and "listen" to what your body wants to do. It may want to remain as it is, turn over, stretch, yawn, sit up, laugh, cry, jump up and run, or do something else. Give in to your body's needs, and enjoy the liberty of the experience until it has run its course.

1-1

Exercise 3:
Stretching the Back Muscles

EXERCISE 3 *(Figure 1-1)*

Sit on the floor with the legs in a comfortable position, the torso relaxed slightly forward, the head hanging down, and the eyes closed. Again go through a mental inventory of where the individual parts of the body are and how they feel. Are you tense anywhere? Is the position uncomfortable? Next, relax the torso slowly forward so that the weight of the head and torso fully stretch the back muscles. When the body has completely relaxed forward, slowly begin to straighten the torso, starting at the base of the spine and working upward to the head. At the end, sit as tall and stretched upward as possible. Imagine that a string attached to the top of your head is pulling you higher. Starting at the base of the spine, again relax by degrees downward to the forward position. Then lift the *head* first, working down to the base of the spine as you sit up in

4

the erect position. Following that, relax downward, starting with the head. As you do the exercise, move slowly and be aware of the sensations of movement, the parts of the spine being used, and your feelings in doing the movement. Repeat the exercise at a slightly increased tempo, breathing in as you rise and exhaling as you relax. Sense or imagine the expansion and relaxation of your lungs in conjunction with your body's expansion and relaxation. Awareness of the expansion-relaxation, inhalation-exhalation sensation is particularly important to the performer, as almost every gesture and all vocal production is related to that concept.

EXERCISE 4

While comfortably seated with your eyes closed, bring the palms very close together without touching them and try to feel the warmth generated between the hands. Next, move the palms slowly past one another a number of times in opposite directions. Move them in a sideward motion, a forward-backward motion, and a circular motion. Then bring the palms close together and separate them. Be aware of the sensation of movement in the fingertips, hands, and lower arms. Next, bring the fingertips of the two hands together so that they are touching very lightly; then lightly tap them together a number of times. Then gently bring the palms together, start rubbing them together slowly, progressively increase the speed and energy of the rubbing, and conclude by separating the palms from one another. Take enough time with each action so that you can feel the sensation of movement. Also try to be aware of any sensation elsewhere in the body, and especially in the area of the spine.

EXERCISE 5 (*Figure 1-2*)

Stand with the legs comfortably separated, the eyes closed, and the arms hanging down at the sides of the body. Move the arms in circular paths around the body. For the first circle of the arms, firmly press the palms against the front of the thighs for six counts; then very slowly lift the arms forward until they are over your head. Conclude the first circle moving the arms toward the back and down to the beginning position. For the second circle reverse the direction: back, over the head and down in front to the beginning position. For the third circle press the palms firmly against the sides of the thighs; then very slowly lift the arms to the sides away from the body until they are over the head. Conclude the third circle by extending the arms sideward across the front of the body as they are lowered to the beginning position. For the fourth circle reverse the direction: sideward across the body in front until the arms are over the head. Conclude the fourth circle by moving the arms

5

sideward away from the body and down to the beginning position. The illustrations show the pattern for one arm only. Do the movements slowly, maintaining awareness of the sensation of movement and particularly with the gravitational pull on the arms as they are lifted and the

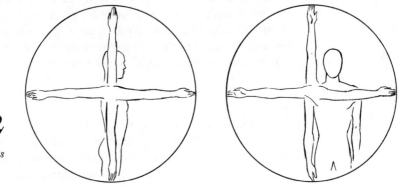

Exercise 5: Arm Movements

release of that pull as they move down to the hanging position. Do the exercise once with a minimum of effort; then repeat it with the fingers spread widely apart and the arms stretched away from the body as far as possible. Do the movement pattern a third time in a rhythmically percussive manner—that is, with quick stops and starts. As you are moving, be aware of the energy required to make the jerky movements, the quality of movement that results, and the feeling in the whole body as you move in this way.

EXERCISE 6

This exercise can be helpful in developing sensitivity to passive and active movements. While standing with the eyes closed, bring the hands to the chest with the elbows lifted to the side, extend the arms in front of the body, bring them back to the chest, and extend them above the head, back to the chest, down beside the body, to the chest again, and to the side at shoulder level. The first time you do the exercise, let the hands and arms be very relaxed. The second time, make tight fists when you bring the hands to the chest, but when you extend the arms away from the body, spread the fingers widely apart. Do the exercise rhythmically, with one count for each movement. The relaxed movements generally use less energy and appear passive, while the stretched actions use more energy and appear strong and vigorous. Repeat the movement pattern a third time, pretending to push something heavy away from the body and then

6

to pull something toward the body. For example, imagine pushing against a wall and then pulling heavily weighted ropes toward yourself. This variation will use great energy and produce considerable tension.

EXERCISE 7

Another body experience that requires controlled relaxation is the shaking exercise. While standing relaxed, let one arm hang loosely from the arm socket. Start by rotating the hand back and forth at the wrist and then slowly involve the rest of the arm. Work for very relaxed and gentle shaking movements. These will require greater relaxation and concentration. Try the same action in other parts of the body, such as the head, torso, and legs. Finally try to get the whole body doing the action in as relaxed a manner as possible.

EXERCISE 8

Imagine a ball of energy centered in the chest. While sitting comfortably, visualize this ball of energy moving slowly and continuously from one part of the body to another. For example, starting in the chest it might move down the arm into and out of each finger, return to the chest, spiral down through the inside of the torso, and then travel through the legs, up the outside of the spine, over the top of the head, down the bridge of the nose, around the eye sockets and the line of the mouth, and back to the chest. Standing up, repeat the exercise, reinforcing the ball-of-energy idea by physically indicating where the ball of energy is in the body at any given moment. For example, if it is traveling from one shoulder to the other and then out through the arm, one shoulder may be lifted, followed by the other shoulder, the elbow, the hand, and the fingertips.

EXERCISE 9

Standing with your eyes closed, touch individual parts of your body first with one index finger and then with the other. For example, touch your nose, ears, hips, knees, and ankles. If you are in an open area clear of obstacles try the same exercise while walking with your eyes closed.

EXERCISE 10

Focusing your eyes on some stationary object, stand on one leg and lift the other leg until the foot is at the level of the knee of the supporting leg. Be aware of your actions and the sensations in your body as you try to remain standing. Relax and then repeat the exercise with your eyes closed. Repeat the exercise several times with your eyes open and then closed. This exercise will help you to understand the importance of sight in helping you to balance against the force of gravity.

EXPERIENCING THE ACTION OF THE MUSCLES

The purpose of the kinesthetic exercises is to help you to experience consciously the sensation of movement. The following exercises should help you experience the sensations of muscle action as the muscles work to move the body from one position to another. Technically you may not be able to actually feel the muscles working, but you can feel them tightening or getting harder. The muscles work as a team. In general one set of muscles does the actual work of moving the body, while other muscle sets act as stabilizers or guides of the action. As one set of muscles shortens in length another set lengthens. By tightening the fist and bringing the lower arm to the upper arm, for example, you can feel the bicep muscles on the palm side of the upper arm shortening. By relaxing the fist and straightening the arm you can feel the biceps lengthening. Ideally every performer should practice physical activity that will not only strengthen and give endurance to the body but also develop coordination and flexibility. Among the many activities valuable for this are dancing, fencing, juggling, gymnastics, and tumbling.

The following exercises cover the major sets of muscles involved in various common movements. Familarizing yourself with the sensations of their action can be helpful in learning the physical skills mentioned above.

Do each exercise slowly five to ten times. Concentrate on discovering whether the muscles are lengthening or shortening and the sensations of the general area of the body involved. Remember that in these exercises you are trying to feel the sensation of muscles working and not trying to build strength, endurance, or flexibility. If you are interested in further exercises, the bibliography provides a list of books that can be helpful in setting up an exercise program for yourself. A few selected exercises are included in Chapter 4.

1-3

Exercise 11: Head and Neck

8

EXERCISE 11 *Head and Neck* *(Figure 1-3)*

While lying on the back, lift the head, touching the chin to the chest, turn the head to the right and to the left, bring it back to the center, and then lower it to the floor. The action involves muscles at the front, back, and sides of the neck.

EXERCISE 12 *Abdominals (Figure 1-4)*

While lying on the back, stretch the arms forward from the chest. Lift the head and then the torso until a tightening in the abdominal muscles between the pelvis and rib cage is felt. Then lower the body to the floor, feeling each part of the spine touch the floor.

1-4

Exercise 12: Abdominal

1-5

Exercise 13: Lower Back

EXERCISE 13 *Lower Back* *(Figure 1-5)*

While sitting on the floor with the legs bent and the arms stretched forward, bend forward at the waist, lengthening and relaxing the muscles of the lower back. Grasp the feet. Return to the original position.

1-6

Exercise 14: Back of Thighs

EXERCISE 14 *Backs of Thighs* *(Figure 1-6)*

Sit on the floor with the arms stretched forward and the legs straight. Bend forward at the waist, relaxing and lengthening the muscles of the lower back and feeling the pull at the back of the thighs. Grasp the feet. Return to the original position.

9

1-7

Exercise 15:
Lower Leg and Foot

EXERCISE 15 *Lower Leg and Foot (Figure 1-7)*
While seated on the floor with the legs straight, pull the feet at the ankle joints back toward the knees until the heels are raised from the floor. Then reverse the action, stretching the foot and toes toward the floor. Try to keep the thighs, lower legs, and feet in one straight line. Be aware of the action of the muscles at the sides and backs of the lower legs.

1-8

Exercise 16:
Arms and Upper Body

EXERCISE 16 *Arms and Upper Body (Figure 1-8)*
Start in a kneeling position with the hands on the floor in front of you so that you are supporting yourself on your hands and knees. Lower and raise your head and torso toward and from the floor by bending and then straightening the arms. This is a modified pushup. Be aware particularly of the action of the muscles of the upper arms and chest.

1-9

Exercise 17:
Sides of the Thighs

EXERCISE 17 *Sides of the Thighs (Figure 1-9)*
Lie on the floor on the left side of your body. Slowly lift the right leg into the air then slowly lower it to the floor. As the leg is raised be aware of the action of the muscles on the outside of the upper leg, and as the leg

10

1-10

Exercise 18a:
Erect Posture

1-11

Exercise 18b:
Basic Alignment

1-12

Exercise 18c:
Front of the Thigh

is lowered be aware of the action on the inside of the upper leg. In order to feel this action better, try the raising and lowering against some kind of resistance. For example, ask a friend to gently push down on your leg as you raise it, and to gently pull up on your leg as you lower it.

EXERCISE 18 *Body Alignment*

a. Erect Posture (Figure 1-10) Placing your back against a wall, stand so that your heels, pelvis, lower back, shoulders, and head are touching the wall. Stretch the arms down toward the floor and at the same time stretch the whole body vertically upward. Starting at the top of the head, try to sense the various areas of the body where the muscles are actively working to help maintain the erect posture.

b. Basic Alignment (Figure 1-11) Step away from the wall. Imagine a line passing vertically through your body, as shown in figure 1-11. Without straining or forcing the action, imagine that your body is being lengthened upward along this imaginary line. The body should have a sense of lightness or floating. Take time to sense how each part of your body, from the top of the head down through the spine, the pelvis, and the legs, is maintained in the basic alignment. Imagine your shoulders as being directed to the sides, so that they are not being forced either forward or backward. Sense your feet in contact with the floor. Imagine the weight as being centered between three points: the big toe, the little toe, and the heel. Most of the major muscle groups of the body are involved in maintaining the erect posture. Try to sense the interplay of the muscles that make it possible for you to stand erect.

This basic body alignment can be thought of as good posture. It is important to you in your daily life to develop the habit of always standing, sitting, walking, or moving with good body alignment and without forcing or straining, in order to conserve energy and to project an image of poise and balance.

c. Front of the Thigh (Figure 1-12) While in the basic alignment position described above, slowly lower yourself toward the floor in a deep knee bend, letting the heels rise from the floor. Do not relax or sit in the deep knee bend and do not bring the upper and lower legs together. Slowly raise the body to the starting position. Be aware of the action of the muscles at the front of the thighs.

In order to avoid injuries, particularly to your knees or ankles, you should always be aware of your body alignment when, supporting your weight on one or both feet, you bend your legs. In this action it is extremely important that you align the center of each knee over the center toes—never over the big or little toe.

11

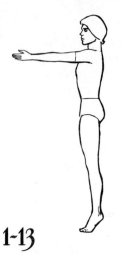

1-13

Exercise 18d:
Back of the Lower Leg

d. Back of the Lower Leg (Figure 1-13) While in the basic alignment position described in section b, slowly rise to the balls of the feet, keeping the knees straight and the body erect. Then slowly lower the heels to the floor. Be aware of the action of the muscles at the backs of the lower legs. As in the knee bend, make sure that the weight is centered over all of the toes and not toward the big or little toe.

EXPERIENCING MOVEMENT DYNAMICALLY

Dynamics is one of the most important aspects of movement for the performer to understand and learn to command. Dynamics is related to how objects move through space. It is concerned with the line of direction travelled, the force behind the movement, the speed of the movement, and the resistance of the moving object to the force of gravity. With human beings changes in dynamics will result in changes in the motivation that causes a person to move; therefore changes in the dynamics will alter the expressiveness of any human movement. Changes in motivation can in turn alter the dynamics of movement.

How you choose to direct your energy in the performance will have a direct bearing on how the audience responds. If you move timidly, the audience will respond, although in a different way than if you move aggressively. If you sit on a chair and repetitiously swing your legs back and forth throughout the performance, the audience may be amused, lulled to sleep, or driven out of the theatre. The more information you can give to the audience through selected use of body dynamics, the more clearly the audience will understand what you are trying to communicate.

Basically all movement utilizes greater tension and lesser tension (relaxation) successively. It is the basic rhythm of life: in breathing, for example, inhaling causes the lungs to fill with air and stretch to a state of tension, while in exhaling the lungs become relaxed. Holding the breath causes body tension; releasing it relaxes the body. Moving the body also involves tension and release. In general as you move, active muscles shorten, resulting in tension, while other "passive" muscle groups elongate, resulting in relaxation. By knowing the degree of tension or relaxation required, you can more quickly visualize and perform a wide variety of movements in accordance with the needs of any particular performance.

The following is a list of actions which clearly show bodily tension and relaxation. Some are characterized by *sustained* tension or relaxation, a *combination* of both, or *alternate* tension and relaxation.

12

1. Pushing and pulling: Pushing and pulling against an imagined or real resistance is characterized by a degree of continual tension.
2. Resisting and yielding to the force of gravity: In standing erect the body resists the pull of gravity. In general the greater the weight the greater the tension. With great tension the body may appear heavy and earthbound, with less tension the body may appear light, and with no tension the body will collapse to the floor. Here the forces to be resisted or yielded to are one's own weight, a part of the body or clothes, and perhaps a mental state inducing a feeling of weight or lightness.
3. Resisting and yielding to forces other than gravity: Resistance is characterized by some degree of tension against a real or imagined outside force, and yielding by relaxation which allows the real or outside force to dominate your movement. In pushing or pulling, as the terms are used above, the performer is considered as active, while in resisting and yielding the performer is considered as acted upon by an outside force.
4. Percussive and sustained action: Percussive actions are sharp, forceful, and repeated, requiring tension to start and stop them. Sustained actions are characterized by smoothness. Sometimes slower movements require more tension for control.
5. Vibratory (shaking) and slash actions: Vibratory actions can be done with tension or relaxation. They are percussive actions with a limited range of motion. The slash action can be done with various degrees of tension. It might be done, for example, with the tension of someone chopping wood or of a conductor leading an orchestra through a very quiet passage of music.
6. Sliding and gliding: Sliding, as used here, implies some outside resistance. Although the action is smooth, some tension is present. Gliding, as used here, implies no outside resistance; therefore the action is smooth and relaxed.
7. Swinging and swaying, with suspension: Swinging a part of the body requires a degree of tension to begin the movement, followed by relaxation as the momentum of the initial energy thrust takes over, followed by suspension (that is, a momentary stop of the motion as gravity takes over), followed by relaxation as the part returns to where it started. Swinging the arms while walking is a common example. The sway is similar to the swing but is more consciously controlled and requires more tension. Both the swing and sway have a natural moment of suspension in them, but suspension can also be introduced consciously into a movement, providing an unexpected dynamic quality. Both

13

of these actions can be done, of course, in any direction and at varying speeds.

To gain a better understanding and mastery of these dynamic contrasts, try simple gestures first, and then move the whole body, as in the following exercises. In all of them, be aware of the amount of force or energy being used, the speed of the action, and the direction which the movement takes through space.

EXERCISE 19 *Pushing and Pulling*

Move the arm forward forcefully, as if commanding someone to halt. Next, using the whole body and the arms, try to push an imaginary heavy object forward. The first action may be done quickly and the second slowly; but both are basic forms of a push.

Pull an imaginary glass of water across the table toward yourself. Then pull an imaginary heavy chair toward yourself.

EXERCISE 20 *Resisting and Yielding to the Force of Gravity*

Stretching the arms over the head and, standing on the balls of the feet, reach upward, trying to touch the ceiling. In the stretched position, release all tension from the body, allowing yourself to fall toward the floor. Repeat the same action with one arm. Start with the arm relaxed at the side of the body. Next, lift the arm with a minimum amount of effort, and then completely relax the arm, letting it fall to the side of your body. Next, imagine that you are wearing shoes weighing twenty pounds each. Standing with the knees bent, try to lift each foot. By contrast, imagine yourself moving in outer space with your body free of the pull of gravity.

EXERCISE 21 *Resisting and Yielding to Forces Other than Gravity*

Imagine that someone holding onto your hand is trying to pull you forward. Pull against the applied force. Next, allow the imaginary person to pull you forward with minimal resistance.

EXERCISE 22 *Percussive and Sustained Actions*

Forcefully strike your fist against a table top. Next, rapidly tap the table top with your index finger, as if you were typing with one finger. Perform the same two actions in slow sustained motion as smoothly as you can. Do them at first with as little tension as possible and then with as much tension as possible.

14

EXERCISE 23 *Vibratory (Shaking) and Slash Actions*

Imagine that you have just washed your hands. Relax the arms as much as possible and shake the excess water from them. Next, imagine that you are holding a baseball bat. Swing the bat to hit the ball.

EXERCISE 24 *Sliding and Gliding*

Place one hand on the top of your other arm (near the shoulder) and press firmly down. Keeping the pressure firm, move the hand down the arm so that a resistance is set up. Repeat the action with the hand lightly touching the arm.

EXERCISE 25 *Swinging and Swaying with Suspension*

Relax the arms, swing them forward to shoulder level, and let them fall of their own weight. Continue the motion, swinging them back. Repeat the movement a number of times.

The sway is similar to the swing but is more controlled. Starting with the feet apart and the body rigid, move the body so as to shift your weight from one foot to the other.

Swing the arms forward over the head, while at the same time rising to the balls of the feet and stretching the whole body upward. Lean slightly forward so the body must eventually fall forward, but resist the action of falling forward. The suspended moment is that moment of balance before the weight of your body pulls you toward the floor.

EXERCISE 26 *Dynamic Changes*

Combine several of the above exercises so that one action flows into another. Use the whole body in doing the action. For example, if you combine pull, slash, light weight, heavy weight, and percussive action, you get a total phrase of movement that has dynamic variety. As a further example, you might extend the arms forward, pull them in to the chest against a great resistance, throw them down and away from your body in a slashing motion, let them rise up above the head as if they were floating upward, quickly drop your whole body toward the floor as if it had turned to iron, and conclude by thrusting both fists forward of the chest as if punching the air. Try this and other combinations a number of times, exploring changes in the speed, space, and energy used.

EXERCISE 27 *Dynamic Gesture*

By investing a gesture with specific dynamic qualities, you can often reveal something about a character's state of mind, basic personality, or age. Take, for example, a basic gesture such as extending your arm forward from the chest with the palm up, as if to indicate to someone that you want him to give something. Try it with different dynamic qualities, each time imagining that the impulse to gesture starts in your chest, travels down the arm, and is concluded when the arm and fingertips are stretched as far forward as possible. The point of this is to assure that the gesture is convincing, and appears full and complete. This gesture can be done in several ways. If the movement is yielding, it might suggest a certain passivity or lack of conviction. If it is then done in a slow and sustained way with a good deal of tension, it might suggest someone in command. Done in a percussive way, it might suggest an impatient demand. Done slowly with a quiet vibratory action, the gesture might suggest timidity, fear, or age. By failing to extend the arm all the way out, you might reinforce the idea of age or indecisiveness.

EXPERIENCING MOVEMENT RHYTHMICALLY

In order to alter the dynamics of a movement, changes in energy and speed are necessary. Movements made against real or imagined resistance will of course be slower, but movements done without resistance will vary in speed according to the demands of the action. For example, in throwing a baseball overhand, the arm swings backward and then is thrust forward with increasing speed to put the maximum energy behind the ball.

Varying the speed of your movements not only adds dynamic variety to a performance but can also help to define characterization. A waitress with only a few customers will probably move more slowly than a waitress serving twenty. On the other hand, the waitress with only a few customers may be so eager to please that she cannot move fast enough, and conversely the waitress with twenty customers may feel that in any event she must put in an eight-hour shift, so that there really is no point in hurrying.

Obviously, given movements are performed in varying lengths of time and rhythms, like music, in which beats combine into measures to form musical phrases. Movements are likewise joined together to form "movement phrases." Speed and rhythm can affect the expressive quality of movement just as they affect the expressive quality of music. The following exercises are presented to help you experience kinesthetically some of the elements of rhythmical movement. Repeat each exercise a number of times until it feels somewhat natural.

To sense the rhythmical pulse of movement, perform exercises 28 through 30.

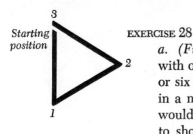

Starting position

3

2

1

1-14

*Exercise 28a:
Conducting Pattern*

1-15

*Exercise 28b:
Leading a Distant Orchestra*

a. (Figure 1-14.) Imagine yourself as the leader of an orchestra and with one arm lead the orchestra in measures made up of three, four, five, or six counts. Use the same amount of time for each beat. For example, in a measure made up of three counts, the arm, starting over the head, would move downward beside the body on the count of one, sideward to shoulder level on the count of two, and to the beginning overhead position on the count of three. Technically, of course, a conductor does not use as much space as indicated here, but to sense the use of space as well as rhythm, exaggerate the actions.

b. (Figure 1-15.) Imagine yourself as the leader of an orchestra seated several blocks away from you. So that your actions can easily be seen by this orchestra, involve your whole body in the actions and cover as much space as possible. Conduct in measures made up of three, four, five, or six counts. The illustration suggests the movement **for** a measure made up of three counts. Use the same amount of time for each beat.

EXERCISE 29

Involving the whole body in the action, stretch both arms over the head on count one, drop the arms toward the floor on count two, and slap the floor on count three. Then, standing up, stretch the right arm to the side on count one, stretch the left arm to the side on count two, and fold the arms into the chest on count three. Do the exercises alternately a number of times so that you can keep the counts at a steady tempo. Each time you repeat the exercise, increase the tempo and be aware of the

17

amount of energy required to move faster and how it affects the dynamics of the movement.

EXERCISE 30

Next, using the same number of counts per measure, move different parts of the body on each new count. For example, using three counts to each new measure, lift the right arm on count one, move the head on count two, and lift the leg on count three; then drop the leg on count one, bend the torso on count two, and drop the right arm on count three, and so forth. Do each movement sharply.

To sense the rhythmical flow of movement, perform exercises 31 through 33.

EXERCISE 31

Continue to count in phrases of three counts. Repeat the actions of the previous sequence, but change the amount of time used for each action. In the previous exercise each action took one count. By doing each separate action in three counts instead of one count and by moving smoothly rather than sharply from position to position, you change the rhythmical flow of the movement. Changes in the speed and flow of movement change the dynamic quality.

For example, lift the right arm in three counts, move the head in three counts, and lift the leg in three counts; then lower the leg in three counts, bend the torso in three counts, and so forth. Do each action slowly the first time through. Try to move smoothly from position to position. Maintain the same speed in counting as you did in the previous exercise.

EXERCISE 32

Continuing to count in phrases of three, repeat the exercise above, lifting the arm in two counts and moving the head in one count, then lifting the leg in one count and lowering the leg in two counts, and then bending the torso in one count and lowering the arm in two counts. The actions should thus be done in speeds of slow, fast; fast, slow; and fast, slow.

EXERCISE 33

Continuing to count in phrases of three, repeat the exercise above with the following variations. Lift the arm in one count, hold the position for one count, and move the head in one count; then lift the leg in one count, lower the leg in one count, and hold the position for one count; then bend the torso in two counts and hold the position for one count; and then lower the arm in two counts and hold for one count. The actions should thus be done in speeds of fast, hold, fast; fast, fast, hold; slow, hold; and slow, hold.

To coordinate two simple rhythms, perform exercises 34 and 35.

EXERCISE 34

Walking in place, keep the feet stepping in phrases of four counts each, each step being done in one count. At the same time move the arms in circular paths around the body in phrases of four counts as follows. Starting with the arms down beside the body, lift the arms to the front and above the head in four counts; move the arms to the back and down to the beginning position in four counts; crossing the arms in front of the body, raise them over the head in four counts; lower the arms away from the body to the sides to the beginning position in four counts; raise them away from the body to the sides and over the head in four counts; and finally lower them to the beginning position by crossing them in front of the body in four counts.

EXERCISE 35 *(Figure 1-16)*

Walking in place and at the same time moving the arms in two different

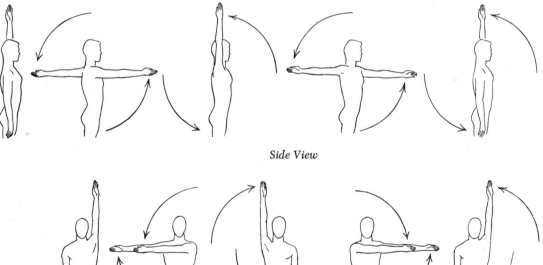

Side View

1-16

Exercise 35:
Arm Movements

Front View

directions is more difficult. Repeat the exercise above with a slight variation in the circular arm movements. Walk with the feet in the same four-count phrase. As the coordination is more difficult, practice the following circular paths with the right arm, then the left arm, then both arms, and finally with both arms while stepping in place. (The figure shows the simultaneous action of both arms.) Start with the right arm down beside the body and lift the arm to the front and above the head in four counts. Move the arm to the back and down to the beginning position in four counts, raise the arm to the side away from the body and above the head in four counts, and finally lower the arm to the beginning position crossing the arm to the left side of the body.

Now start with the left arm extended over the head. Move the arm to the back and down to the hanging position in four counts, lift the arm to the front and above the head to its beginning position in four counts, lower the arm to the hanging position by crossing the arm to the right side of the body in four counts, finally lift the arm to the side away from the body and over the head to its beginning position in four counts. Repeat the exercise a number of times until you can keep the speed of the steps constant while smoothly coordinating the movement of the arms in the two opposite directions.

You can develop some awareness of the rhythm and phrasing of movement by watching others move—for example, in a restaurant. Try to assign a general metrical pattern to the movement you watch. For example, a waitress comes up to someone's table and offers him a menu which he takes, opens, and reads. Try to assign so many counts for each action. The customer looks up at the waitress (two counts), acknowledges her and the offered menu (two counts), and reaches for the menu and brings it to himself (two counts). In total this could be called a six-count phrase. It has a beginning, a middle, and an end. The waitress sets up an event to which the customer must react. He must make the decision to take or reject the menu. He does so, thus concluding that phrase of movement. The important point in this kind of analysis is to develop awareness of a simple phrase of movement and to become aware of the natural time elements for its individual actions. They can then be consciously lengthened or shortened to help illustrate a state of mind or characterization. Timing will be altered by such considerations as how secure the customer feels in the restaurant and toward the waitress, how hungry or thirsty he is, the shape and size of the menu, and the atmosphere of the restaurant.

You can develop a sense of rhythm by setting up simple dramatic scenes of this kind for yourself and then performing the basic actions in varying rhythms for various emotional states and characterizations.

Other good sources for observing movement phrases and rhythms are sports activities and animated cartoons. Televised sports shows often feature slow-motion replays of individual actions. Thus you can see a whole phrase of movement, including the initial impulse, the way the athlete moves through space, and how the movement is terminated. It is important to learn to see the impulse, the follow-through, and the completion of a movement so that, when you yourself move, your movements are realistically motivated and look complete to an observer. In cartoon shows the action of the animated figures is precisely timed so that with a few simple movements a series of cartoon gags is put across. Watching cartoons can help develop your appreciation of the importance of timing or rhythm in movement.

It is also helpful to learn to sense the rhythm of your own movements. For example, the next time you perform an action such as brushing something off your clothes, off a table, or off a chair, analyze how you did the brushing. How fast was the action? How much tension was involved? Was the action smooth or sharp? Did you use the palm side of the hand or the back side of the hand? Did you use only the fingertips? Among other actions you might observe yourself doing are scratching, rubbing, waving goodbye, picking up something, writing, turning the pages of a book, and washing your face. Almost any action is a valuable source for analyzing the mechanics of movement and how it is affected by motivation.

EXPERIENCING MOVEMENT SPATIALLY

As a performer you should explore the space around yourself and become familiar and comfortable with it. The *kinesphere concept* is useful in exploring the space around yourself while standing in one place. Rudolf Laban defined a person's kinesphere as the spatial area around the body in which one could move either the whole body or a part of the body.

EXERCISE 36 *(Figure 1-17)*

To understand this concept imagine yourself in the center of a cube with lines radiating out from your body in every conceivable direction. Then imagine the various individual parts of your body as placed on the same spatial line as any one of these imaginary radiating lines. As a physical exercise, place one arm in line with these various imaginary radiating lines. Let the body twist or bend as necessary in order to get the arm in every possible position around the entire body.

Next, try the exercise with one leg and then with other parts of the body, such as the elbow, the hip, or the torso. Try exploring your indi-

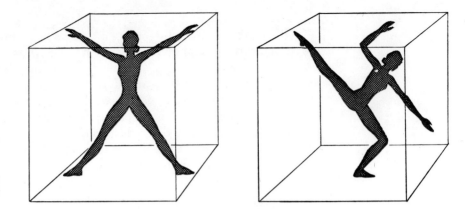

1-17

Exercise 36:
Kinesphere Concept

vidual kinesphere while lying down, kneeling, standing, and moving around the room. Each time you move to a new area in the room, think of yourself as surrounded by your kinesphere.

EXERCISE 37 *(Figure 1-18)*

Select from a magazine or newspaper three or four pictures that show individuals in different poses: for example, lying down, sitting, kneeling, bending over, throwing something, dancing, modeling clothes, or playing sports. Also try to find pictures that indicate how the people must have felt when the pictures were taken: for example, sitting in deep thought, throwing something in anger, kneeling in worship, or dancing with joy.

1-18

Exercise 37:
Spatial Designs

Imagine that each person is in Laban's kinesphere, and with a colored pen or pencil trace the most obvious lines of design his body makes in space. Next, try to pose yourself in a way similar to each picture. Next, move from pose to pose a number of times until you can make the movement transitions smoothly, and then try to reverse the movements from the last pose to the first pose, as if a film were being

22

run backwards. Finally, while moving from pose to pose, try to assume the feeling expressed in each picture you selected. Repeat the whole sequence a number of times until it flows smoothly from one position to the next and feels comfortable to you. Then try the same actions varying the speed with which you move from pose to pose.

When these various static positions are connected, one has movement. It is similar in principle to still movie frames. Each frame is an arrested movement frozen in space, but when the film is run through the projector the individual positions become animated with motion.

In moving from pose to pose with the whole body or any of its parts, you are creating a visual pattern in space. In one sense, there are really only two basic kinds of patterns you can trace: patterns of straight lines or curved lines. With straight and curved lines combined, all kinds of spatial patterns can be formed: for example, triangles, squares, circles, figure eights, and wavy lines.

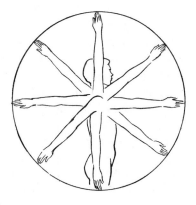

1-19

Exercise 38:
Movement Patterns

EXERCISE 38 *(Figure 1-19)*

To experience the kinesthetic sense of your movement patterns, extend the arm forward of the chest and imagine holding a piece of chalk in your hand. Take the arm over the head, to the back, down beside the body, and to the front. Had you been writing, you would have made a circle. Next, starting with the arm extended forward from the chest, bring the hand to the chest, then extend it to the side in line with the shoulder, and then return it to the front with the arm straight. The pattern of two straight lines and a curved line makes a modified triangle.

EXERCISE 39

As an exercise to explore the space around yourself and to develop a sense of the spatial pattern being made, imagine a piece of chalk attached in succession to various parts of your body, such as the tip of the toe, the knee, the elbow, the left hip, or the nose, and then in space draw patterns such as squares, rectangles, figure eights, letters of the alphabet, and words. In order to really use the body and all of the space around it, make the movements broad and big.

A person's use of space may reveal the kind of person he is or at least his present mental state. For example, an extroverted, confident, outgoing person might use the space around himself in a different way than an introverted, inward-looking, cautious person.

If you intended to establish an extroverted character, his movements might follow paths that were broad and expansive in the use of space, away from the body, and pulled vertically upward, giving an illusion of lightness and freedom. The introverted character, on the other hand, might make movements that were restricted, close to the body, and pulled vertically downward, giving an illusion of tightness and slowness.

EXERCISE 40

In order to get a sense of how different people use space, start with the concept of the extroverted or introverted person suggested above. Set up a series of natural movement patterns that one does every day and do them as the introverted or extroverted person might do them if there were a number of strangers in the room. For example, walk into the room, be introduced to someone in the room, sit down, stand up, pick up something from the floor, open a window, and turn around to see who has just entered the room. Next, select a few characteristic postures and movement patterns that suggest the kind of person you were portraying. Arrange the postures and movement patterns in sequence and repeat them a number of times until you feel natural doing them. Repeat them, being conscious of the visual lines that your body is creating in space. Repeat the sequence again, being aware of how the movements affect your own feelings. If your movements are restricted or free, does that alter your own feelings in any way? How fast or slow are your movements? How much tension is present in the movements? What parts of your body are you using and how are you using them?

EXPERIENCING THE MOVEMENT POSSIBILITIES OF THE INDIVIDUAL PARTS OF THE BODY

It is important that as a performer you become aware of exactly where each individual part of your body is placed and the overall silhouette your body is creating in space. It is also important for you to know the movement possibilities of each individual part of your body. It is the

24

small details of placement and movement of the individual parts of the body that give subtlety or shading to the basic silhouette. Attention to detail distinguishes an accomplished performer from a beginner.

In analyzing the positions that parts of the body can take, it is helpful to categorize the movements as follows:

1. Movements in a specific direction away from or toward the body.
2. Movements to some level in space.
3. Movements with or without rotation or twisting.
4. Movements lengthening or shortening a part of the body.

For example, the arm can be extended forward. It can be at a level straight forward of the chest. It can be rotated outward in the shoulder socket so that the palm is up. Finally, it can be stretched or lengthened varying distances from the body or bent toward the body.

To get a better understanding of how your body moves, you can determine which movement categories apply to each of its parts. For purposes of analysis, the body parts are the head, shoulders, rib cage, pelvis, whole torso, arms, legs, feet, and hands.

Certain movements may have emotional or rational implications. For example, the head can rock forward and backward as if saying "Yes," rock side to side as if in a jaunty mood, or twist side to side as if saying "No." The head can thrust forward as if in aggression, move back as if in repulsion, or move sideways as if covertly straining to overhear a conversation. The neck can stretch with pride or contract into the shoulders with timidity (figure 1-20). The shoulders can move up and down as in a

1-20

Head Positions

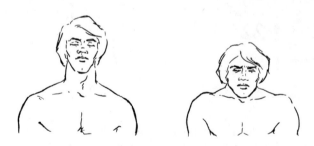

shrug, forward as if the character is shy, or back as if proud. They also can twist back and forth—perhaps the action of a coquettish woman. The rib cage can move forward with aggression or back in suspicion. It can twist side to side along with the shoulders. It can be sunken, suggesting an introverted character, or it can be lifted and expanded, suggesting a confident person. The rotation of the pelvis forward (tucked under) usually denotes an extroverted character while rotation backward (swaybacked) may suggest introversion.

25

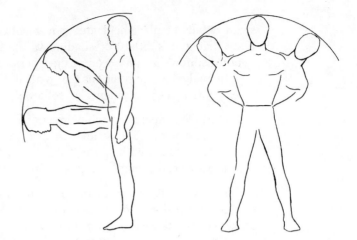

1-21

Torso Positions

Whole torso actions can be broader than the foregoing (figure 1-21). The torso can bend forward at the waist as in bows, bend back as was done by the women in the early Renaissance, and bend sideways, perhaps to support oneself against a wall. Like the head, shoulders, pelvis, and rib cage, the torso can be rotated from side to side or pushed as a unit in some direction (figure 1-22). It can also be contracted inward (concave)

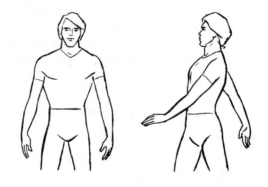

1-22

Further Torso Positions

in a manner of total introversion or stretched upward and lifted (convex) as in total self-confidence and extroversion.

The legs and arms have greater freedom of movement than the other parts of the body but can be categorized in the same way. Carriage and placement of the arms and legs is very important, for example, in period plays or dances where fashionable carriage implies good breeding. The slightest change of direction, level, or rotation of an arm or leg can make the difference between an awkward or graceful position, which is why it is important to observe carefully distinctions of direction, levels, and

26

1-23

Arm Design

rotation. Figure 1-23 indicates an extreme change of arm design by rotation in the arm socket.

The whole body can be lowered toward the floor or raised upward on the ball of the foot. It can leave the floor, as when jumping. The whole body can also turn to the right or left, making a partial or complete revolution while supported on one foot or on both feet, or while in the air.

EXERCISE 41

In order to develop coordination and learn to move the various parts of the body independently, you should explore the various possibilities for movement indicated above. Setting up a rhythm for yourself, start with the head and move it rhythmically in all possible directions in isolation and in combination with other movements. Then, working downward to the other parts of the body, repeat the isolated actions and combinations. If you can work with someone, have him clap a rhythm for you and shout out the new part of the body which you should move at any given time. Working with someone in this way can be helpful to you in learning quickly to identify, isolate, and use various parts of the body.

EXERCISE 42

a. While standing, keep the arm and hand in a straight line and move them around the body. The range of movement for the arm is quite great, although there is some natural restriction in movements to the back.

b. Bend the lower arm and move the whole arm around the body in different ways. You should be able to get the bent arm into areas around the body where you could not when the arm was straight.

c. While bending and rotating the lower arm, rotate the upper arm in the shoulder socket inward and outward and observe how the direction of the lower arm changes with changes in the upper arm rotation. Observe where the palm faces with changes in the rotation of the lower arm.

EXERCISE 43

This is a very useful exercise to explore the variety of arm and hand positions. Take a handkerchief between the first two fingers, let it hang freely downward, extend the arm to the side at shoulder level, and do the following:

a. Without moving the upper or lower arm, move the hand around in different ways.

b. Keeping the upper arm in place but allowing it to rotate in the shoulder

27

socket, move the lower arm and hand into different positions. Be aware of the changes of line and the effect of the action on the handkerchief.

c. Moving the upper and lower arm, wrist, and hand, explore the movement possibilities. Where it helps the movement, involve the head in the action. Include motions that bring the handkerchief to the waist, the shoulder, the nose, and the forehead. These actions could be useful in period plays.

EXERCISE 44

We rarely use the movement potential of the hand and fingers. Explore some of the many possibilities of the whole hand, as follows.

a. Hold the fingers straight and together. Stretch them backwards so that the hand is arched. Then curl the fingers forward so the hand forms a cup. Perform the same exercise with the fingers separated.

b. Starting at the fingertips, fold the fingers in toward the palm of the hand so that the fingertips rest on the edge of the palm. Then fold the fingers in toward the palm so the fingertips rest as close to the wrist as possible. Finally fold the fingers into the center of the palm to form a fist.

c. Try folding each individual finger toward the palm while keeping the other fingers straight. Is it possible?

d. Try moving each finger individually in a sideward motion and then a forward and backward motion.

e. Move each finger individually in a circular path.

EXERCISE 45

Try the same movements with the whole leg, lower leg, and foot that you performed with the arm and hand. The leg and foot actions are naturally more restricted. In order to develop some sense of the unexplored potential of the foot and toes while barefoot, try to pick up various objects, such as a shoe, a marble, a small ball, a pencil, and a piece of paper. Try to hold a pencil between your toes and write on a piece of paper. With practice your writing could become quite legible, but the purpose here is only to explore the use of the feet.

*. . . craft is not only for the externals of acting,
but for the internals . . . this is a craft
in which one can train oneself.*

LEE STRASBERG

2 Exploring Movement

THE SENSES

Contact with our environment is made through the five senses: sight,
smell, taste, touch, and hearing. These five senses communicate the exterior world to the brain. In recent years the sense of touch has been
further catagorized into the four primary skin sensations of cold, warm,
pain, and pressure, and into the organic sensations, which are kinesthetic,
static, and visceral. The latter three provide information to a person
about his bodily activity and about his movement, balance and body
position.

How well a person uses the senses depends in part on his heredity,
environment, and sometimes special interests such as, for example, music
or painting.

As a performer you will communicate a great deal to an audience on the nonverbal or sensory level. For that reason sensory awareness must become "second nature" to you through conscious exercises that use and develop your senses. Your body and mind should be trained to distinguish subtle differences in sensory quality and to utilize these differences before an audience. A successful performance for both you and the audience will be partly dependent on the skills you acquire. For our purposes, a perceived sensation has four attributes: quality (distinctive attributes), intensity, extensity (size in space), and duration.

THE TRADITIONAL FIVE SENSES

Touch is the sensation conveyed through the tactile nerves by pressure, traction, or temperature applied on the skin or mucous membranes. One can distinguish between temperatures ranging from hot to cold, between degrees of pressure ranging from light to heavy or painful, between moisture and dryness, and between rough and smooth textures. One can also distinguish shapes and sizes by touch. Related to touch are the kinesthetic, static, and visceral senses. Through them we are aware of our bodily movement and posture; we are aware of our weight and the muscles, tendons, and joints through which we execute movement; and we are aware of our balance when still or moving. In addition, visceral sensations lead to the awareness of such things as hunger.

The sensations of *taste* and *smell* are related. Man can distinguish only four basic tastes on the surface of the tongue: sweet, primarily on the tip of the tongue; sour, at the sides of the tongue; salty, around all the edges of the tongue; and bitter, at the back of the tongue. Smell is generally defined in terms of its intensity as scent, odor, or aroma. Apparently man can distinguish six basic qualities of smell: fruity, flowery, resinous, spicy, foul, and burnt. Temperature, pressure, taste-smell relationships, and the chemical sensitivity of the individual also affect the perceived sensation.

Sight is basically the transmittal of visual stimuli to the brain through the eyes and nervous system. Man can perceive light, distinguish among colors on the basis of their various wavelengths, and perceive shapes. To do this he must separate an object from its background. Through vision man is also able to orient himself in space, by distinguishing distance, movement, and the direction of movement.

Hearing and sound production are of course related. The outer ear receives sound waves, the middle ear conducts them to the inner ear, and the inner ear translates them into patterns of nerve impulses which are sent to the brain, resulting in auditory perception. In man the auditory experience can be described in terms of the pitch, loudness, and timbre of what is heard. The slower the repetition of the frequencies of the wave pattern, the lower the tone. Loudness is the product of the intensity of the wavelength and its frequency. The timbre is the quality of a tone; it is the timbre that is the characteristic quality of various human voices and musical instruments, for example.

To perform more effectively you must learn again to experience *yourself* as a living entity. Like others, you are doubtless drawn into activities that take you away from yourself, so that it is important to re-establish contact through your senses with yourself and with the seeming "silence" around you. Sensitivity training exercises, which have become popular in recent years, can be beneficial to you in this way.

The following exercises should be performed in an almost ritualistic way, with concentration and in silence. Their purpose here is not therapeutic but to help you to rediscover your own body and the material world through your senses. The exercises deal with four kinds of experience: eating, drinking, contacting yourself, and contacting nature. You may wish to broaden the range of experience in other ways.

EXERCISE 46

Choose one or several simple foodstuffs, such as a single nut, a raisin, a grape, a bite-sized piece of bread, a lemon, a carrot, a stalk of celery, or a piece of lettuce. Look at it and "relate" to it. Examine its texture, color, shape and size. Smell it. Next, with proper respect, eat a small portion of it. What is its texture? What are the sounds as you bite into it and chew it or let it dissolve in your mouth? Be aware of how it feels as you swallow it. Next, try the same exercise with a liquid, such as water, wine, coffee, tea, or milk. Examine carefully the vessel that holds the liquid and then touch the liquid to experience its temperature and its density. Smell the liquid and study its color and the hues reflected in it. When you drink, hold some of the liquid in your mouth to experience the texture and taste before you feel it slide down inside of you. You might try the same thing while blindfolded.

The same kind of experience can be had the next time you eat a meal by yourself. Look at the food on your plate and study its color, size, and texture. As you eat each bite, really experience the texture and taste of the food and the sound or silence as you move the food around in your mouth. Then feel the food moving down into your stomach.

EXERCISE 47

This exercise is to help you to make contact with yourself as a biologically functioning human being. Lie or sit down comfortably and in

31

silence. You may wish to place your fingers gently over your closed eyes, or your hands over your ears or your heart. Become aware of your breath inhalation and exhalation. Listen for any sounds your body makes. Become aware of where you are tense or relaxed. Be aware of where different parts of your body are and how they are positioned. Be aware of pains or nervous reactions in various parts of your body. And finally be aware of *yourself being aware of your body* and be aware of *your reaction to your awareness.*

EXERCISE 48

Still lying down, run your hands over your body, outlining the various shapes, contours, and textures. Imagine them in your mind, and later study them in a mirror or try to draw them on paper. You should also smell and taste your clothes and your body.

As you remain quiet, listen intently to the "silence." What sounds do you hear? How does the body feel in contact with the object supporting it? What smells are present? If your eyes are shut, what are the lights or colors and shapes passing before your eyes? Is there any taste in your mouth?

EXERCISE 49

Another rewarding experience is to take a walk somewhere specifically for the purpose of using the senses. The walk can be almost anywhere: beside the ocean, in the mountains, in a park, in a market, or on a city street. Again the idea is to use fully each of the senses. You might ask yourself such exploratory questions as: How many colors are there in the sky, on the ground, and in people's clothing? How do the people, trees, clouds, and cars move? What are the shapes, sizes, and textures of the objects? What are the different sound qualities heard and how does one contrast to another? How does the wind (or the sun, the rain, or your clothing) feel as it contacts your body? Are there smells present and do they invoke images of taste or something else?

It is important for you as a performer to awaken all your senses to the experience of living. Consciously try to identify the natural odors of people, the colognes used to cover their natural odor, the odor of a woman as opposed to that of a man or a baby, the odors of different houses (where the odor of food or household pets, for example, may be strong), and the odors of certain sections of your city or town.

In your daily living begin to explore your sense of touch by touching different fabric textures, hand soaps, hand creams, book covers, water, oils, woods, sandpaper, glass and china, rocks, leaves, and other objects.

32

Try touching parts of your body with various objects, such as the point of a pin, a strand of hair, a feather, and the eraser end of a pencil. Apply varying pressure. Move various objects across your skin, such as a kleenex, a piece of paper, silk, a blade of grass, and a hand towel. Try dipping your hands first in cold and then in hot water, and then dipping one hand in cold water and the other in hot water at the same time.

Explore the capabilities of your sense of sight by becoming aware of such things as the play of light and shadow on objects; subtle differences in color; the many different colors in the sky, grass, and trees; the relationships between objects arranged next to one another; the various kinds of printing used in advertising; the sizes and shapes of leaves and the way they are arranged on a plant or tree; and the movements of insects.

Keep your sense of hearing alive by listening carefully to the voices of other people: not only what they say but how they say it. How do they pronounce words? How fast do they talk? What inflections do they use? What qualities of voice do they have? Listen to the different sound qualities which airplanes, automobiles, people, and animals have when they are close and when they are far away. Listen to the sound of water running, of bacon frying, of water boiling, and of a fly in flight. Other valuable exercises are those involving recall of sensory experiences.

EXERCISE 50 *Recall of Sight*

a. Think of a room with which you are familiar. Try to describe in detail every object in the room. Describe the colors, textures, shapes, sizes, and smells. If there are books in the room, try to describe their order on the shelf, the titles and authors, and the colors and designs of the jackets. Does the printing on some of them have an odor?

b. Describe in detail a friend you have seen recently. What was the friend wearing? What color are his eyes and hair?

c. Watch a group activity involving, for example, people moving around, birds flying, or fish swimming. Mentally trace the patterns the subjects make as they move in and out of your vision. Then close your eyes and try to visualize the same movement patterns as completely in detail as you can.

d. At a respectful distance, follow someone who is walking, and then try to mimic every detail of his walk. Later try walking the same way.

e. Watch your friends carefully and notice how they move, sit, stand, and gesture, and in general the timing of their movements. Later try to

33

move the same way. If your friends are good-natured about such things, try doing your experiment in front of mutual friends to see if they recognize whom you are imitating.

EXERCISE 51 *Recall of Touch*

a. Act out the experience of walking with bare feet on hot sand at the beach or on hot pavement.

b. Act out taking a shower. Turn on the water, test the water temperature (too hot) with your hand, and adjust the water to a moderate temperature. Take the shower, soaping up and rinsing off, bracing yourself for a last rinse off with very cold water. Finally dry yourself off with a rough towel.

c. Notice an imaginary button missing from your clothing, get out an imaginary thread and needle, thread the needle, and finally sew on the button. Try the same action again, characterizing someone with bad eyesight, someone inexperienced with needle and thread, and someone who is very nervous. Be alert to the size and weight of the imaginary objects you are handling, the kind of fabric you are sewing, and the length of the thread.

d. Pick up from a table several imaginary objects of different weight, size, and shape. Your movements and the way you handle each object should reveal very clearly what you are doing. Useful objects are a teacup, an apple, a knife to cut the apple, a telephone, a book, and a pencil.

EXERCISE 52 *Recall of Smell and Taste*

Pick up an imaginary object that has a distinctive odor or taste and act out your reaction to it. A few possibilities are picking and then smelling a flower; opening a jar of dill pickles and then smelling and tasting one; slicing a lemon and sucking on it; tapping, rolling, perhaps wetting, and lighting up a cigar and then smelling the smoke after exhalation; opening a package of limburger cheese and perhaps reluctantly having a taste; and eating a spoonful of honey or peanut butter. Be particularly aware with each new smell or taste what happens to your breathing pattern, your mouth, lips, and nose, and the skin around your nose, eyes, and forehead. As a final exercise, go through the mechanics of eating a meal in pantomime, carefully handling the utensils, reacting to the smell and taste, chewing according to the texture of the food, and finally making it clear whether the food is difficult or easy to swallow.

EXERCISE 53 *Recall of Hearing*

a. Dramatize a recent conversation you have had with someone, using

34

"I said" and "he said." Try to reproduce the conversation in as exact a manner as possible. Consider the word order, the sentence order, the rhythm of the speech, the pronunciation, and the mood of the speaker. Repeat the exercise, either imitating the voice of the speaker, substituting some other voice characterization (for example, the voice of an old person, a braggart, or an introvert), or using a regional accent.

b. Act out your reaction to various sounds: for example, a knock at the door very late at night, an unexpected harsh and painful sound such as a low-flying jet airplane, or music that is particularly appealing to you. In this last example, react appropriately as each new musical instrument plays—the drums, the trumpets, and the violins.

c. Try to imitate many kinds of sounds: mechanical or musical sounds, for example, and those of nature and animals.

USING IMAGINARY PROPS

In a performance, you may be called upon to use actual props or to pantomime their use. Aside from helping you to master the art of pantomime itself, pantomimic exercises will aid you in dealing effectively with real props on stage. They will teach you what actions you really perform when handling everyday objects and point up elements of physical characterization which may otherwise escape your notice and which will enrich and add conviction to your performance. For this reason the pantomime required by the exercises below should be as realistic as possible and not stylized. The test is clarity. If someone watches your pantomime and cannot tell exactly what you are doing, then you probably have not conveyed your intentions with sufficient precision of movement.

In doing the following exercises, perform the suggested actions first with a real prop whenever possible and then in pantomime. Try to establish and maintain throughout the pantomimed action the actual size, shape, weight, and texture of the object being used. See how many different ways you can use each prop. Use the prop as a young or old person, as a man or woman, as a member of one social class and then another, and under the influence of differing states of mind. Be sure to consider the use of force, rhythm, and space in each pantomime. To get a better feeling of its weight, texture and shape, you might handle each prop while blindfolded.

EXERCISE 54

a. See how many different ways you can lift, hold, and use a cup, glass, pencil, fork, spoon, handkerchief, book, newspaper, magazine, flower, and other objects.

b. Lift and drink from a cup, glass, bowl, and spoon.

c. Drink hot liquids such as coffee, tea, milk, and soup. Drink cold liquids such as milk, water, Coca Cola, brandy, and champagne.

d. Lift a book from the floor as if it weighed from one to thirty pounds. Next, try the same action as if you were lifting barbells, a cat, an infant, a bag of groceries, a foot stool, a large trash can, an empty suitcase, and a full suitcase.

e. Turning a door knob, open the door, exit through the doorway, and then close the door behind you. The first time open the door toward you and the second time away from you. Be aware of the size of the door knob, the action of turning the door knob, the effort required to open and close the door, and where your body must be in order for you to let the door swing past you. Then open and close windows that open toward you, away from you, and upward and downward. Alter the timing and dynamics for characterization.

f. Dress and undress. Be aware of the cut and fabric of the clothing and buttons or zippers. Include in your clothing such accessories as scarves, ties, belts, handkerchiefs, and hats.

g. Put on stage makeup, give yourself a permanent, or cut your fingernails and toenails.

h. Prepare your own cigarettes or pipe for a smoke.

i. Haggle over the price of something that you eventually buy and that requires an exchange of both bills and coins.

j. At a standup buffet dinner, pick up a glass, a plate of food, a napkin, and silverware, and try then to figure out how to eat while still standing.

k. Think of activities you might do around the house and do them in pantomime. For example, prepare the sink to wash dishes, wash them, dry them, and put them away; or sweep a linoleum floor and a rug; or dust furniture, a knick-knack shelf, and other objects.

l. Pantomime other activities: for example, repeatedly throwing a ball into the air and catching it; flipping a coin into the air and catching it; taking a key out and unlocking a door; picking up a telephone receiver and dialing a number; opening an envelope and removing the letter; opening a drawer, taking something out, and closing the drawer; or turning on a tap to fill a glass with water.

36

Invent a silent scene that requires imaginary props and perhaps other imaginary characters. Keep in mind that you must not only reveal the size, weight, and shape of each object but that you must also make clear your character's attitude toward the objects and other characters. For example, when you are picking up a glass, make sure that your hand does not become a fist. If you are talking to an imaginary person, make his height and location clear at all times.

An example of a pantomimic scene is the making of a "Dagwood sandwich." You could go to the refrigerator and take out such things as bread, cheeses, lunch meats, relishes, and spreads. You could pantomime the actions of slicing cheese, smelling the relish, tasting the meat, licking the spread from each finger, and building the sandwich. You might then, with great anticipation of the event, delay the pleasure of eating it by wringing the hands together, rubbing the fingers or palms together, opening and closing the fingers as you prepare to pick up the sandwich, and then deciding that you had better put one more item in it. Finally, as you lift the sandwich to take a bite, the center could fall out and spill down your front.

Other scenes that you may wish to elaborate on are described below.

a. A cowboy from the Old West comes into a saloon, orders a bottle and a shot glass, and then proceeds to pour and drink glass after glass of whiskey.

b. You have recently had by telephone an anonymous threat made on your life. A friend with whom you have recently had a very serious misunderstanding has just brought you a beautifully wrapped present. How will you handle the package?

c. You are alone in a remote cabin in the mountains during a snowstorm and have been reading a popular mystery story. There is a knock on the door. After looking out the window, you open the door, but no one is there. As you prepare to sit down again, you think you hear footsteps in the hallway upstairs. You light a candle and go to investigate.

d. You are the strongman in a carnival sideshow. You lift heavy weights for the crowd. After the crowds have left you pick up the weights, which are made of plaster of paris, in order to store them until the next show.

e. You are in a room without furniture or windows. The only door has been locked from the outside. You have a strange feeling that the room is getting smaller. You then become convinced of it, and the walls and ceiling do begin to come closer and closer together.

f. You are waiting on a busy street corner for someone who is late. You watch and react to the people going by you. By your posture, walk, or gestures you make clear exactly who is passing by you.

g. You are a blind man walking down the street with a cane. You pass by other people walking down the street and come to the curb of a busy intersection.

h. It is winter. You are without friends or money and sitting alone in an apartment without heat. The landlady comes in and presents you with a bill for your overdue rent.

EMOTIONS

It is sometimes said that although a person may lie with words, his body will not lie. For example, children, who have not learned to hide their bodily expressions, have difficulty understanding how their parents know when they are not being truthful. Although adults are usually more skillful at masking their true feelings, careful observation of them can often reveal how they actually feel. For example, a person who is smiling pleasantly and attempting to appear at ease may nevertheless reveal his unease by pressing his hands together or wringing them. If the bodily movement contradicts the facial expression, the face may not be a true reflection of the person's feelings.

Even the position of the body may reveal a person's feelings toward someone else. For example, Julius Fast in *Body Language* says that if a person is "shutting out" someone, he may cross his legs or arms *away* from that person and also turn his body slightly away. If he is "open" to the person, he may turn toward the person or cross his arms or legs in his direction. Physical expressions of feeling can generally be interpreted quite accurately.

Tests have shown, for example, that people can identify an emotion accurately by looking at a picture of a person's face. When the pictures included the body posture, however, the accuracy of identification was improved.

The importance of nonverbal emotional communication between you and the audience cannot be overemphasized. Some people, in fact, respond more fully to emotional messages conveyed by body movements than they do to verbal communication. It has already been noted that physical expression can bring unique depth to a characterization. Thus it is equally important that you learn some of the ways in which specific emotions are expressed physically and gain experience in coordinating these expressions for the purpose of subtle and convincing characterization.

Charles Darwin, in *The Expression of Emotions in Man and Animals*

(1896), and Carl George Lange and William James, in *The Emotions* (1922), made early and important contributions to the understanding of emotions from a physiological and psychological standpoint. Darwin observed that actions which we associate with the expression of particular emotions are closely allied with the functions of our nervous system and are often performed independent of will and habit. In other words, Darwin said that it is often impossible to experience a particular emotion without giving it physical expression.

Darwin also observed that certain movements which have become habitual in the expression of specific emotions also manifest themselves involuntarily when a situation arises that is *similar* to the habitual stimulus. For example, we may unconsciously make a grimace of pain when someone describes the pain he has experienced. Further, *recollection* of some sensory experience (such as the tasting of a lemon) may cause the body muscles and nerves to respond as they did to the original experience.

Darwin and James and Lange stated that there were four basic emotions that could be accurately identified by their physical manifestations: joy, grief, fear, and anger, as shown in figure 2-1.

2-1

Bodily Expression of Emotion

The emotion of joy is manifested physically by some or all of the following signs. The whole body is drawn upward and appears light, buoyant, and youthful. Movements are swift, alert and rhythmical. Various random and purposeless gestures and sounds may be produced, such as clapping, jumping, stamping, laughing, or the squeals of delight so often performed by children. Decisions to act are made quickly. Talk is rapid and fluent, and is usually louder and at a higher pitch than normal. The face looks rounder as the cheeks are drawn upwards. The lips are curved upward in a smile and the brow is unwrinkled. The eyes are bright. The skin is tinged with color.

If laughter follows it is characterized by deep inhalation followed by quick repeated spasmodic contractions of the diaphragm. The body may shake and a swaying of the head or body back and forth may result. If the laughter is intense, the body may be thrown backwards. The mouth may be open to varying degrees, the lips drawn upward, the eyebrows slightly contracted and the eyelids partially lowered. Tears may flow easily.

The emotion of grief may be expressed in several ways, depending on the circumstances. *Frantic* grief, for example, might be expressed by someone upon hearing the news that a loved one had died. It is characterized by excessive emotional energy and fits of body tension. Such actions as wild pacing of the room, tearing at the hair or clothing, wringing of the hands, and random jerky gestures that touch the head or the body may be observed. The physical manifestations of *deep prolonged* sorrow are in direct contrast to those of joy. The whole body appears drawn downward, weighted down, and weakened. The head hangs down. The person may be motionless or may make only minimal movements, which are slow and languid. There may be passive rocking back and forth or occasional cold shivers. The person leans on things for support. When walking, he drags the feet and holds the arms limply at the sides. The facial features become elongated and take on a sunken appearance as the eyelids, mouth, and jaw all relax downward. The forehead is contracted and the eyes are dull, expressionless, and heavy-lidded. The color is drained from the face. The person may remain silent and "lost in thought." Deep sighs are drawn. When he speaks the pace is usually slow, the volume soft, and the pitch low. Prolonged sorrow may lead to total prostration. The person may seek relief in tears. The face may be reddened and swollen. Sobbing is produced in a manner similar to laughing, with a deep inhalation followed by sharp, repeated contractions of the diaphragm.

The physical expression of fear also differs with the intensity and duration of the emotion. Initially there may be a deep inhalation of breath, a freezing of the muscles in tension, and perhaps a cold shiver. The muscles are ready for whatever action is required, whether it be to fight or to run. The arms may be extended above the head. The hands may cover the face or head. The hands, with the palms turned outward from the body, may be extended toward the object feared as if to ward off a blow. The mouth is open. The eyebrows are raised, and the eyes are wide open and staring, with eye movements being made from side to side. If the cause of the fear is not immediately apparent, the body

40

may remain in this state of motionless tension to listen and see. The breathing may stop or at most continue quietly. A sudden exhalation of breath and relaxing of the muscles may follow the initial intake. Other characteristics are loss of color, trembling, sweating, erection of the hair, rapid breathing, loss of muscle control, disturbed mental faculties, and convulsive movements of the body or lips. Prostration or fainting may follow. There is generally difficulty in speaking, and the person may stutter.

The signs of anger are also modified by degree. In extreme rage, the body is charged with violent energy, which may be used to threaten, to yell, to strike objects or people, or to stamp the feet. The uncontrolled movements are quick, forceful, clumsy, and generally frantic. The fists are generally clenched shut and the body is either erect or hunched over as if ready to fight. The face is contorted in a frown and the teeth may be clenched or ground together. The lips may be parted, showing the teeth. The veins, particularly around the forehead, neck, and hands, are swollen. The face can be either red or pale. The eyes are open wide and glaring. Exhalation is strong. Breathing is irregular and labored, with an obvious fall and rise of the chest, dilated and quivering nostrils, and possibly a trembling body. There may be difficulty in speaking. The voice is usually harsh and loud.

Anger and indignation are more controlled than rage, but as feelings they are of the same kind. They may be revealed by heavier breathing, clenched fists, and similar facial gestures. In defiance or contempt one side of the lip may be raised to show the canine tooth in a sneer.

Darwin describes some of the physical manifestations of other strong reactions. Pain, for example, is revealed in the facial expression, in the respiration pattern, and finally in writhings of the whole body. Severe pain can cause deep depression, fainting, convulsions, and prostration. In pain the muscles are strongly active; the mouth, for example, may be firmly closed or the lips may be pulled back, exposing clenched or grinding teeth. The eyes are wide and staring, the brows are contracted, and the nostrils may be dilated and quivering. The person perspires.

Most other emotions can be thought of as variants of joy, sorrow, terror, or anger. Thus if you are familiar with the extreme manifestations of these basic emotions, you have a selection of characteristics and levels from which to choose in building a scene or character. Under the general category of joy might be listed such pleasure-giving emotions and attitudes as: desire, faith, love, enthusiasm, hope, romance, ecstacy, euphoria, spirituality, and happiness. These are generally characterized by a feeling of lightness in the body, erect carriage, free-flowing movements, and a generally pleasant demeanor. Grief is allied with melancholy, sorrow, and depression: the body feels heavy and seems to pull inward and downward

41

on itself. Fear is related to such feelings as anxiety, terror, shock, panic, and surprise. These feelings may cause a sudden convulsive movement and extreme body tension. Mere surprise may subsequently evoke laughter, whereas strong fear may produce violent and uncontrolled actions of self-protection. Anger is related to jealousy, hatred, revenge, greed, superstition, rage, annoyance, and suspicion. These result in body tension and may produce convulsive movements such as twitching of the hands or facial muscles. The angry person may even beat on some object as a release for tension.

A performer must, however, make his own observations. It is his perceptions that help to make his enactment unique and convincing. From the beginning, learn to question your own feelings when analyzing a character. How did I feel in an analogous situation? What caused me to feel that way? How did I express my feelings? Did I show my feelings or did I try to hide them? Did I make movements with my face, hands, or body that revealed my feelings despite my effort to conceal them? If I gave free expression to my feelings, what movements did I make? When I felt an emotion, what was the sequence of feeling from inside my body until the emotion was expressed externally with movements? Similar questions can be asked when you are observing others.

By careful observation of others you can develop understanding of body language and learn to use it to express character. For example, watch a waitress to see how she greets customers not only verbally but with her eyes and mouth. Is she happy to serve them, or is she angry at the cook, tense because of the rush hour, nervous because she is new on the job, or scowling because her feet hurt? The inventive performer develops this kind of awareness and "files" his observations away for use in the later development of a character. Nothing is lost: all such experiences will reappear when they are needed. The following exercises are intended to increase your powers of observation and analysis.

EXERCISE 56 *Observing Expressions of Emotions*

a. In a restaurant, observe other groups and individuals talking and eating. Concentrate on the physical manifestations of their reactions and try to label them as to mood and emotion; if you cannot hear the conversation, so much the better. If the mood is one of levity, try to analyze the body actions leading to the release of laughter and the subsequent calming down. If the mood is more serious, what kind of body postures are the diners assuming? Are they tense? Are they leaning close to one another or are they physically withdrawn? This kind of exercise can also be done by turning the television set on and watching the picture without the sound.

42

b. Find a place where young children are playing and observe them without their awareness of it. Carefully note how they express their feelings physically and with sound. Listen to their conversation for what they say, how they say it, and their physical reactions to what is said. Be especially aware of their emotional changes. If you are unable to watch children, watch the actions of adults in supermarkets, stores, and other places where they are required to interact with one another. Their emotional-physical actions may not be as clearly defined as those of children, but they are often expressed in more subtle ways. Television is an additional source for observing others. Particularly valuable are such shows as documentaries, news programs, talk shows, and audience-participation shows.

EXERCISE 57 *Observing Your Own Emotions*

Select a television show that has some interest for you. Occasionally when you get involved emotionally in the events of the show, stop yourself and ask what kind of emotion you were feeling, how your body felt, and what kind of body positions or tensions you were assuming as a reaction to the scene. After practicing this kind of analysis for a while, try to analyze the development of the emotion in yourself as it unfolds. If the scene were comic you might note, for example, that you began your expression of joy with a slight turning up of the corners of the mouth and continued with laughter and finally a release of energy as the laughter subsided and you sank back in your chair.

EXERCISE 58 *Recreating Emotional Reactions in Yourself*

What is called the James-Lange theory is useful for the performer. Basically the theory states that if the body is placed in characteristic postures or follows the characteristic actions of a particular emotion the emotion can be aroused. In developing a characterization your previous observations of the physical expression of emotion can be a springboard to an in-depth analysis of how your character feels and expresses himself. By understanding your character's background and function in the play, his emotional makeup and how various emotions can be conveyed physically, you can begin to feel and move as the character might do.

a. Reread the description of the four basic emotions (joy, grief, anger, and fear) and then experiment with each one to see if you can recreate something of the emotion in yourself by moving in a manner similar to the description. At first try each reaction in a sitting and then a standing position. Next, try walking in a manner that suggests the emotion. Finally use imaginary situations to evoke your emotional response. Some useful situations are suggested here. (1) You have just received the news that you have won a two-week vacation to your favorite vacation spot (joy). (2) You have just received news that a favorite pet has passed away (grief). (3) You have just received news that vandals have ruined something that you particularly liked (anger). (4) You have just received news that a terrible fire is raging out of control and that you have very little time in which to escape (terror).

b. A few descriptions of movements that suggest a certain emotional state follow. Although such movements may not always mean the same thing, try moving as the description indicates to get a feeling in your body for the kind of emotion suggested.

Devotion and reverence are often characterized by raised face and eyes and sometimes by kneeling with clasped hands. Reflection and meditation are often characterized by downcast eyes. The person may frown if perplexed by a difficult thought. A meditative state is suggested by a vacant look in the eyes and a squint, as if one were trying to see something in the distance. The head may be dropped forward. Trying to remember something is characterized by a raising of the eyebrows. Determination is suggested by a firm closure of the mouth. Disgust is related particularly to the senses of taste and smell: a wrinkling of the nose as if smelling something bad, or an opening of the mouth as if to regurgitate, often with a gutteral sound being emitted. Contempt is indicated by averting the partially closed eyes or the body from the scorned object, a slight lift of one side of the upper lip, and perhaps a derisive smile or laugh. Slyness is indicated by holding the head still and moving the eyes from side to side. Pride is shown by an exaggeratedly erect body. The eyes are lowered as if looking down on someone, and usually the mouth is firmly closed. In shame or shyness the eyes may be averted and cast downward or sideward and generally are unable to rest in any one place.

EXERCISE 59 *Recalling Personal Emotional Reactions*

Try to recall and describe all the details associated with a strong emotional reaction you have experienced. Include what happened, who was present, and where and when it happened. Try to recall the sensations you felt at the time and how you moved physically in response to them. Next, try to pantomime the actions and gestures you might have made. Finally repeat the pantomime, trying now to simulate your original emotion.

EXERCISE 60 *Establishing Character Physically*

With the help of a close friend who is not aware of what you are doing, try to express physically some of the characteristics of an emotion such as grief or joy. Note whether your performance elicits a comment or reaction. If not, note whether your friend begins to take on the physical characteristics associated with your emotion. If you are projecting the emotion strongly enough your partner should react in some way. After the experiment tell him what you were doing and why you were doing it and then discuss with him his reactions.

DEVELOPING MOVEMENT FOR A CHARACTER

The following are selected methods to establish a proper way of moving for any character being portrayed. In the beginning it is usually preferable to exaggerate rather than to understate or repress your natural expressive inclinations. Many directors agree that it is easier to subdue the performer than to energize him and arouse him to spontaneous gesture. In the beginning, explore freely in every way the expressive physical attributes of a character. Later, through thoughtful selection, a number of the distinctive attributes can be utilized to reveal the character externally. Those attributes which you have felt internally have a much greater chance of arousing emotional and intellectual reaction from the audience.

Following are some exercises to help you to establish kinesthetically various emotional states.

EXERCISE 61

Think of some occasion in your life to which you responded with very strong emotions. The actual event may be significant or insignificant: for example, when someone crowded in front of you in a line. The stronger your reaction was, however, the easier it may be to recollect. Sit quietly and try to remember every detail of the event and try to react emotionally as you did then, but *without* actually making any gestures. Try to be sensitive to how your body is reacting under the stress of "reliving" the experience. If you have successfully called up the experience, there should be some kind of cyclical development of the emotion from its inception to its climax and remission.

EXERCISE 62

To use the cycle of the emotional expression you have discovered, stand

45

up and, using the whole body, try to exaggerate the kind of feeling you experienced when sitting quietly. For example, suppose that you chose the emotion of anger. The possible cycle of physical manifestations might be a rush of blood to the extremities and head, followed by great tension, an impulse to strike out, and a release of energy that leaves the body weak. To exaggerate this feeling in the body you might slowly lift the arms to the side while breathing in and imagine that your whole body is swelling to enormous proportions. Then as the body tenses you might do a deep knee bend while pulling yourself into a tight ball. As the urge to strike out takes over, you might quickly stand up and fling your arms out, feeling as if your body had exploded in every direction. This might be followed by a complete relaxation that leaves you with the feeling that your legs are too weak to hold you up. The point of the exercise is to exaggerate the physical manifestations of the emotion beyond the realm of reality.

Do not allow yourself to feel inhibited in the exaggeration of the emotion. Society conditions people to control their emotions in every situation. What you as a performer want to do, however, is to open yourself to the way it feels to move freely in space in an expressive way.

EXERCISE 63

Using the same emotion as above, select three or four actions typical of such a feeling. Again, if the emotion were anger, you might choose (1) a slow inhalation of air while lengthening the spine, (2) a slight leaning forward of the torso with the head thrust forward and the fists clenched, and (3) picking up a book and banging it down on a table. Physically combine the three actions together. The first few times go through the actions very slowly with very little energy and tension. Complete each movement fully before you go on to the next one and be aware of how the movement feels in your body. Repeat the sequence slowly again, using exaggerated energy and tension in the body. Repeat the sequence slowly again and imagine that you have been offended by someone as you lengthen the spine, that you are directing your anger toward that person as you lean forward, and that you are striking out at the person as you bang the book on the table. Repeat the sequence a number of times at increased speeds and be aware of the effect of the speed on the body and in the expression of the emotion. Repeat the sequence while speaking a few sentences that express your feelings.

It is sometimes more dramatic to make absolutely no gestures but rather to sit quietly and intently internalize the cycle of the emotion as you did in the first exercise in this chapter. If you internalize it very

46

strongly, the audience may pick up your projection. At other times one simple well-chosen gesture will be more effective. For example, you might sit quietly and internalize the emotion and at some appropriate point strongly press your fist into the palm of your hand to show both your anger and the fact that it is being held in check.

When you have analyzed the text of a scene and feel that a certain state of mind is called for, this kind of preliminary exploration can help you discover and internalize the gestural patterns that fit the scene. There is always a danger that deciding arbitrarily on a certain gesture in a certain place will produce a mechanical effect because the gesture has not been internalized. This kind of preliminary exploration may be helpful in avoiding that.

THE FACE

On stage or off, people rarely use the full potential of their facial gestures for communication. Society conditions us to use only certain facial muscles to produce "acceptable" expressions: the mild frown of discomfort, the pleasant smile of minor satisfaction, and the expressionless mask indicating "sophistication." In acting, however, the face should be used as fully as the voice and body to express a character's state of the mind.

On film, where the camera can enlarge the face, the facial gestures are vitally important, but on the stage, where the performer may be separated from the audience by some distance, facial gesture may seem less important from the communicative standpoint. They are, however, important even from a distance because of their effect on the whole body, which in turn helps in communicating with the audience.

Following are some general observations and exercises to develop control and awareness of the face's expressive function both in silence and with speech.

EXERCISE 64

Looking into a mirror, explore the movement possibilities of one part of your face. Imagine movements for that part and at first move the face with your fingers to help the muscles to follow your directions. Keep trying until you get the facial muscles to follow your original concept. Do not be concerned with whether an expression represents an emotion or some other part of the face is forced to move also. Be concerned only with exploring what that single part of the face can do. Explore it until

47

you have a feeling for what muscles control the movement and until you can repeat it at will. Keep in mind that your ultimate goal is the command of your own body, including your face, for the purpose of characterization.

If you decide to move the eyes first, remember that they can move up, down, sidewards, or in combination movements. They may cross as they look down the nose. The eyelids may be wide open or tightly closed. They may be singly winked. They may be slowly lowered or rapidly blinked. The eyebrows may be drawn together, raised, lowered, or raised and lowered in quick succession, either separately or together. The ears may be wiggled while raising a single eyebrow or may be moved independently. The nose may be pulled up as if smelling something unpleasant, or contracted downward, narrowing the nostrils. The nostrils may also be flared laterally, as horses sometimes do. The lips may be curled down in various "frown" positions, or lifted up in various "smile" positions. They may be pulled laterally as if one is smug or self-satisfied. One side of the upper lip may be lifted in a sneer, wrinkling the nose and exposing the teeth. The upper lip as a whole may be lifted, similar to the movement rabbits make. The lips may be pinched together, making the cheeks hollow. One side of the lip may be diagonally pulled up while the other side is pulled down to form an oblique line, or one side only may be pulled up or down.

The mouth may be wide open, with the teeth either exposed or covered. The tongue may be extended from the mouth to either side or up or down. It may be run over the lips or over the teeth. The teeth can bite down on the tongue, or the tongue can be extended in a contemptuous gesture.

EXERCISE 65

After exploring some of the possibilities for isolated features, try combining these facial gestures. Then try to build up expressions representing specific emotions. Following this, involve the whole head in the expression, then the shoulders, then the torso, and finally the whole body.

EXERCISE 66

After exploring the movement possibilities of the face and integrating them into expression of the whole body, try to determine the expressive values of each facial movement as you make it. Does raising the eyebrows express surprise or a sense of superiority? Does lowering the eyebrows express displeasure of some kind or does it merely denote thinking? Do

48

downcast eyes suggest shame or thought? If the eyes are moved sideward and downward does it suggest contempt or shyness? If the mouth is opened wide is it expressing fear or great joy? If the teeth are firmly clenched and the lips pressed tightly together, does this express determination, defiance, or anger?

In exploring the face, you have no doubt noticed the lines created in the face by the use of certain muscles. As the aging process takes place these lines will become deeper. Recognizing some of these basic lines and observing them in older people, you should be able without the aid of makeup to simulate to some extent the facial features of older people or of a particular character type. Jean-Louis Barrault, the brilliant French actor and director, said that scenery was necessary in the modern theatre because the actors were not able to provide it. To some extent this is true of the use of makeup: it should be used only when the performer cannot create and sustain a complete illusion with his own body and face.

EXERCISE 67

a. Masks can also help you to perform new body movements. They are often available in novelty stores or toy stores. If you can get some interesting masks, put each one on in front of a mirror and perform movements in the style of the character suggested by the mask. For example, try picking up a cup to drink, examining something, or seeing someone come into the room.

b. Try to establish different kinds of masks *with your own face.* Standing in front of a mirror, imagine some kind of distinctive face, hold your hand over your face while you are changing your expression, and then reveal the new face to the mirror. Do this a number of times, changing your face, for example, to that of a clown, an angry person, a bitter person, a person without teeth, a person with bad eyesight, a prudish person, and a suave person.

EXERCISE 68

A great deal can be said with the eyes. Awareness of eye focus is very important, as it can add immeasurably to your expressiveness in performance. The following exercises will increase your understanding of the expressive use of the eyes.

a. Walk across the room as if you were walking down a street. Imagine that someone is coming toward you from the opposite direction. See him from a distance, make eye contact with him before you pass him, and look away as you go by.

49

b. Repeat the walk several times and with your eyes establish the person as very short, the same height as yourself, and then taller than yourself.

c. Repeat the walk again, establishing with your eyes a definite emotional reaction to the other person. For example, establish that you feel friendly, you feel unfriendly, you feel frightened or uneasy, your are intrigued, you flirt, or you recognize the person as a friend.

d. Sitting down, imagine someone in the far distance coming toward you until he is within a few feet of you. Be aware of widening or narrowing of the eyelids and the kind of strain placed on the eyes during observation.

e. Place an imaginary pencil on a table. Look at the eraser on the pencil, lift the pencil, and bring it slowly to within an inch of your eyes. Notice the way your eyes and the area around them react.

Further Exercises

Several other exercises can be performed with a partner. For example, the two of you can sit opposite one another and try to match each other's expressions, no matter how exaggerated. Later you may try affecting certain expressions and ask your partner to tell you what feelings he thinks they represent. The following exercise, used in some psychological programs, is also valuable for increasing your powers of observation and characterization. Sit opposite your partner and symbolically either lift his head from his body and place it on your own shoulders or lift your head and place it on his shoulders. If you both concentrate sufficiently, the one receiving the head should undergo a clearly definable change of facial expression and body attitude to reflect those of the person across from him. Even if the exercise is not totally successful, careful scrutiny of the other's face will in itself offer beneficial practice in observation.

3 Visualizing Movement

You can often express the state of mind or the dominating characteristics of a given character with a few well-chosen body positions or movements; therefore it is very important that you practice visualizing and coordinating static and moving body designs in order that they can be clearly seen by an audience, whose understanding in a theatrical production is quite often arrived at through what it sees rather than through what it hears. Just as an actor explores the possibilities of his voice and speech patterns, he should do likewise with his postures and movements.

There are three things to consider: the static design the body makes in space, the moving design the body makes in space, and the expressiveness of the stance, posture, and gestures.

The body is a three-dimensional form in space. The shapes the body

51

makes in space consist of straight and curved lines. For example, if you drop your head forward and round the spine from the top of the head to the pelvis, the torso design is curved. If the arm is stretched away from the body at shoulder level, the line of the arm is straight.

The basic design your body creates in space may be symmetrical or asymmetrical (figure 3-1). In symmetrical design both halves of your body

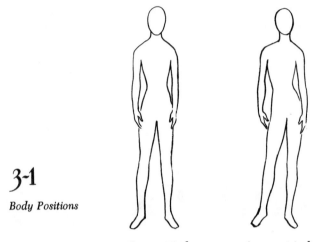

3-1

Body Positions

Symmetrical   *Asymmetrical*

are placed in space in exactly the same way. For example, if one arm were raised to the side at shoulder level then the other arm would be raised to the same level and held exactly the same way. In asymmetrical design the halves of your body create different shapes. For example, if one hand were placed on the hip and the other were placed behind the head, your body would be in asymmetrical design.

The body in space also suggests shapes *around* itself (figure 3-2). The shape of the body in space could be called the primary design, and the shapes around the body could be called the secondary designs. Both contribute to the overall visual and emotional effect.

In order to visualize the spatial designs created by the body, it is helpful to define "normal" stance as a reference point. In "normal" stance the body is held upright in a good posture, the arms are relaxed and hanging down at the sides, and the feet are placed together with the toes pointed forward.

When the body is upright in a good posture and supported by the feet, it is balanced on its own center of gravity. The center of gravity is an imaginary point in which all the parts of the body are balanced in relationship to one another and the point where the primary weight of

3-2

Primary and Secondary Designs

the body is concentrated. In the above upright posture this would be the pelvic area. The line of gravity is an imaginary vertical line that always passes through the center of gravity (figure 3-3). Changes in posture or, for example, carrying a heavy suitcase would cause the center of gravity to change.

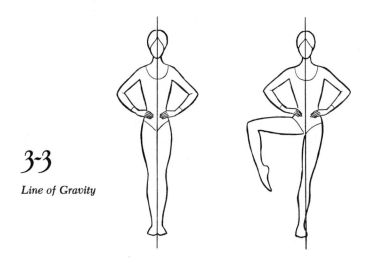

3-3

Line of Gravity

Any shift or movement of the whole body or any of its individual parts requires a compensation or readjustment of the body in relation to the line of gravity. When the body is not balanced on center of gravity and balance cannot be recovered, the body will fall down. In an action like walking, balance is momentarily lost but is immediately recovered. The placement of the center of gravity is important for the performer to master in case he is asked, for example, to stand on one foot or to faint on the stage. Such mastery also provides for freedom of action, complex body coordination, and a general sense of stage grace.

To sense your own center of gravity, assume different positions: for example, stand on one leg with the other foot lifted as high as the knee of the supporting leg. Then move the free leg to the front, back, and side. Next, lean forward at the waist, then back and to the side. Try the same positions while holding heavy books at arm's length in front of you. In each case try to determine how your body shifts away from the original line of gravity and where the center of weight is concentrated.

The body can be viewed from six basic viewpoints: front, back, either side, the top of the head, and the soles of the feet. When you observe someone, you may first be aware of his face, then the color of his hair or

Anteroposterior

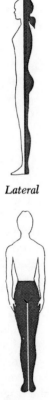

Lateral

Transverse

3-4

Planes of Motion

eyes, and later the shape of his body or other distinguishing characteristics. Later you may have the opportunity to view him from different vantage points. All such views will contribute to your total impression of him. Without straining credibility, a performer should try to be seen at various times from different viewpoints.

Using these six viewpoints, move your whole body or a part of it in one or more directions. For example, while standing, move front, back, to either side, toward or away from the floor, and diagonally.

In order to analyze movements from a visual standpoint, you should know some of the concepts commonly used in analyzing motion. For purposes of analysis the body is divided in three planes of motion which correspond to the three dimensions of space: height (up-down), width (side-side) and depth (forward-backward). The important fourth dimension is time, of course. A plane of motion can be visualized as a sheet of glass slicing through the body in one of these three dimensions (figure 3-4). The vertical, front-back plane (anteroposterior) slices through the body from front to back, dividing it into right and left halves. The vertical, side-side plane (lateral) slices through the body from side to side, dividing it into front and back halves. The horizontal, upper-lower plane (transverse) slices through the body, dividing it into upper and lower halves.

Movements of the body forward or backward are in the front-back (anteroposterior) plane of motion: for example, bending forward at the waist or lifting the leg backward. Movements of the body sideward are in the side-side (lateral) plane of motion: for example, bending sideward at the waist or lifting the leg sideward. Rotary (twisting) movements of the body around the longitudinal axis of the body are in the upper-lower (transverse) plane of motion: for example, rotating the head to the side. It is possible also to move in two planes of motion at once: for example, a forward bend of the torso midway between the front and the side.

In analyzing the moving or the static design of the body in space you can determine for each: (1) the direction moved in one of these planes, (2) the distance moved away from the previously defined "normal stance," (3) whether the visual line is curved or straight, (4) whether the design is symmetrical or asymmetrical, and (5) whatever additional details are present.

BASIC POSTURES

A person reveals a great deal about his attitude and general disposition by the way he stands, sits, walks, and gestures. A consideration of some

54

basic positions and movements and their expressive qualities is helpful in learning to plan movement. It must be remembered that postures and gestures should be motivated from within to avoid the appearance of artificiality.

For purposes of analysis *stance* is defined as one's basic position: for example, standing straight with the feet together or apart. *Posture* is the spatial relationship between the various parts of the body. *Gesture* may involve the whole body but here it is defined as the movement of an isolated body part, such as a shift of the eyes or raising of the arms.

While standing or sitting, the body creates a silhouette which can be altered by arm and hand placement, carriage of the shoulders, rotation of the leg in the hip socket, basic stance, carriage of the head, and the angle of the body toward the audience. Basically the silhouette can be seen as a series of straight or curved lines.

A first body posture is one that is vertically lifted upward (figure 3.5).

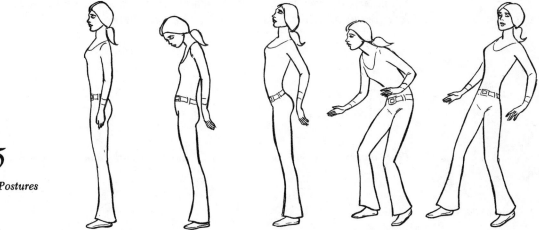

3-5

Body Postures

First Posture   Second Posture   Third Posture   Fourth Posture   Fifth Posture

It is associated with "good posture," in which the base of support (the feet) and all the skeletal parts above are held in alignment. The attitude is expressive of confidence and poise. The body is ready for free movement in any direction required.

A second body posture is a curved line, contracted downward on the line of gravity. The body is concave or rounded in the front. The head is dropped forward, the shoulders forward, and the pelvis rotated under. The attitude is expressive of deep thought, depression, or withdrawal. Mobility is restricted.

A third body posture is a line curved in the opposite direction. It is

55

similar to an exaggerated military posture, with a sway back. The head lifted and tilted slightly back. The shoulders are pulled back, the chest is lifted and puffed out, and the pelvis is rotated backward. This posture is expressive of pride, arrogance, over-confidence, or contempt.

A fourth postural possibility is a tilt forward from the waist to form an acute angle with the rest of the body, with the head thrust forward as if one were trying to see something at a distance. The back may be straight or stooped. This pose could suggest old age, a position of defense, or preparation for an attack. Freedom of movement is somewhat more restricted in the stoop.

A fifth body posture is a leaning back from the waist, with the torso in a straight line. It could be expressive of haughtiness, pride, contempt, defiance, or withdrawal from something unpleasant.

BASIC FOOT POSITIONS

While a performer is standing or seated, some common positions of the feet (based on ballet positions) are as follows (figure 3-6).

First Position   Second Position   Third Position   Fourth Position   Fifth Position

3-6

Positions of the Feet

Feet Pointing Outward   Feet Pointing Inward

First position, feet together
Second position, feet apart, side
Third position, feet together, heel to instep
Fourth position, feet apart, front to back
Fifth position, feet together, heel to toe

The first and second positions may be the most natural positions in which to stand, particularly for men. The fourth position is used naturally in walking, and the static third, fourth, and fifth positions are more likely to be used in period plays by both men and women or in a contemporary

56

play by a woman characterized as a "refined lady." In a contemporary play, the feet are more often pointing forward in a parallel position or slightly turned out from the hip sockets. In certain period plays they are more likely to be turned outward from the hip sockets. Someone from the lower classes, a comic figure, or a shy character might turn the feet inward from the hip socket.

When the weight is equally divided between the feet, the emotional attitude may appear less decisive or committed. It is in effect at rest. If in an erect body the weight is shifted toward the forward foot or to one side, the body position indicates the character's interest in what is happening in that direction. When the weight is placed on the back foot, the body tends to lean slightly in that direction and indicates a withdrawn, uninvolved, or superior attitude. When the weight is evenly divided between the two feet, dynamic energy can be maintained and enhanced by a slight lift of the heels, which also helps to keep the body alert in order to maintain balance.

BASIC ARM POSITIONS

Although there is more variety in the positions the arms and hands can take, the basic positions are similar to those established for the legs (figure 3-7).

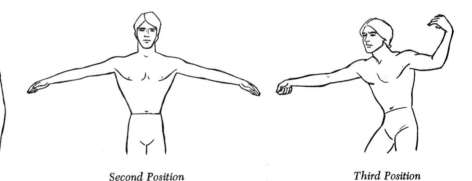

First Position   *Second Position*   *Third Position*

3-7
Arm Positions

A first basic arm position is with the arms freely hanging down at rest beside the body.

A second basic arm position is with the arms extended to the side at shoulder level. The arm can be rotated in the shoulder socket so the palm is facing forward, backward, down, or up. The upper or lower arms can be bent at various angles or extended into a straight line. The hand can be an extension of the lower arm or it can drop forward at the wrist or be pulled backward or to one side. The important consideration about the basic second position is its use in plays taking place from the sixteenth through the eighteenth century. Proper carriage of the body became synonymous with elegance and proper court deportment. A graceful curved arm was extended to the side when dancing or walking. For this, the second basic arm position requires a gentle slope downward from the shoulder to the fingertips. The arm is slightly bent at the elbow. The upper and lower arm are rotated slightly forward so that the elbow looks supported rather than dropped. When working with a partner, the lady's lower arm is often rotated inward so that the palm is down and the man's lower arm is rotated outward so that the palm is up. The lady's extended hand may be gently placed on the man's extended hand. In a period play, there is generally no holding or gripping of fingers or hands.

A third basic position is similar to the forward-backward relationship of the legs, with one arm extended forward and the other back, either bent or straight. This position might be used in sword fighting or in a period bow.

Other positions might be taken with one or both arms extended to some level forward or backward, and either bent or straight.

For further analysis, consider all the ways you can move one or both arms so that the hands contact or relate to the body. For example, such positions could include hands on the waist; hands supporting the chin or head; hands joined behind the neck; hands supported on the legs; hands joined in front of the face, at the chest, at the waist, or in the lap; the fingers or hand scratching, rubbing, massaging, or pulling on some part of the body; the index finger pointing toward some part of the body; biting of the nails or cuticles; and other gestures bringing objects to the mouth to taste, to the nose to smell, to the eyes to see more clearly, and to the ear to hear better. The possibilities are endless.

SITTING IN A CHAIR

Sitting is affected by the style and size of one's chair. Figure 3-8 shows a few basic postures. For example, an erect posture with the feet directly in front of the chair and the legs reasonably close together suggests a refined character or a formal situation. A less formal or detached attitude is suggested by relaxing the body, stretching the legs, and draping the arms casually on the chair.

When you are seated, various parts of your body can assume many positions. For example, the arms can be folded on the lap, wrapped around the waist, placed on the knees, draped over the chair, or spread out to the sides. The torso can be erect, with the hips far back in the chair or on

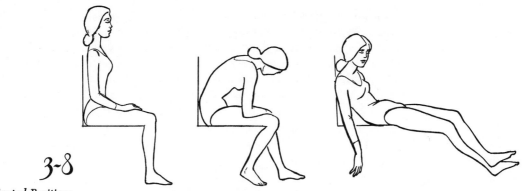

3-8

Seated Positions

the edge of the chair. The body can be slumped down or erect on the edge of the chair. The torso can be tilted forward and supported by the hands, arms, or some object like a table or cane. The legs can be spread apart, turned in or out from the hip socket, or kept parallel. They can be hooked around the chair legs, extended out in front, or crossed at the ankles, or one leg can be crossed over the other at the knee. The feet can be tucked under the chair or placed in front of it. They may rest flat on the floor, on their sides, or on the balls or heels.

The young and the less inhibited frequently relax totally into a chair, spreading their legs or tucking them under in some way. The more formal tend to sit erect with feet together. Men sit in various ways according to their habits, background, or profession, but generally they are able to sit more informally than women because of custom and clothing. Women in dresses usually but not invariably keep their knees together and may cross their legs at the ankles. With advancing age the feet begin to turn outward when resting on the floor. In old age men tend to keep the knees together, while women's knees are separated, the legs apart.

In sitting down there are three general approaches based on age. The young tend to throw or drop themselves into a chair, while older people lower themselves into a chair with or without the help of their arms. As age increases and the body joints become stiffer, people lean forward and lower themselves into a chair with the support of their arms. These procedures are also dependent on individual attitude. You should practice all of them.

One of the most difficult things for most beginning performers is to sit in a chair with a minimum of bodily adjustment. The technique is relatively simple and should be practiced until it feels natural. A basic sitting technique for the stage is to walk to a chair, finishing with one foot very close to the far corner of the front of the chair, make a half-turn pivot on that foot so that the edge of the chair can be felt with the back of the leg, and then, keeping the torso relatively straight, lower oneself into the chair. If it is done correctly, the hips will be well back in the chair. To stand up, reverse the action. Pull one leg back so that the weight is on the ball of the foot, and move the other foot slightly forward with the weight on the whole foot. Keeping the torso straight, use the back foot to push yourself up, ending with all of the weight on the front foot. You are then free to step out on the back foot.

If you are already standing in front of the chair with your back to it, the technique for sitting down is the same. Move one leg slightly back to feel the edge of the chair with the back of your knee and then lower yourself. If you choose to sit down with arm support, you can face the chair, hold onto its back, arms, or edge, and then without letting go of it turn and lower yourself. A second approach is to face away from the chair, reach back to hold onto some part of it, and then lower yourself.

THE FEET

It is amazing how little we consciously use our feet in expression. We usually do not point out things with them, most people do not use them to pick up things, and we rarely use them to touch someone affectionately. Even in dance they are not really exploited to the extent to which the hands are.

The feet are perhaps less expressive when supporting the body than when we are seated, but in either case they can still reveal some inner states of mind. A person may reveal his inner state of mind, for example, by toeing the ground, stamping or grinding out a cigarette violently, rubbing his feet against the carpet or grinding his heels into it, tapping his toes against the ground, stamping his feet impatiently or shifting from one foot to the other, wiggling his toes with pleasure, rubbing one foot against the other foot or leg sensuously, or placing his feet on furniture either to relax or as an unconscious sign of hostility toward the owners of the furniture.

THE HANDS

Tension may be revealed by clenching the hands into fists or rubbing them together; rubbing a finger and the thumb together; rubbing the hands against some part of the body; twisting one's clothing, a ring, or a handkerchief; pulling at the clothing, the hair or some part of the body; or drumming the fingers against something.

60

Even various fingers have become associated with various mental states. The index finger is called the mental finger. We use it to point out something or to touch our forehead, the mind. The middle finger is considered the physical finger. Its use is usually more awkward, earthy, or even vulgar. The fourth finger is considered the finger of affection. It is the ring finger in marriage. The little finger is the finger of affectation, artificiality, or humor. Someone trying to be cultured while drinking from a cup may stick the little finger out. The thumb is considered the most vital of the five fingers. According to Ted Shawn in *Every Little Movement,* François Delsarte found that the thumbs moved toward the palm in death, in ill health, or when a person was not sincere in his expressive gestures. Hiding the thumb in the palm, he felt, was a sign of depression or extreme fatigue. Conversely a healthy or sincere person's thumb, he said, was extended away from the hand when making an expressive gesture. To avoid artificiality while gesturing with the hands on stage, the validity of these definitions should be tested under many situations.

GESTURE

Gesture can be thought of as a silent language, capable of communicating ideas, sentiments, and attitudes. Some gestures are made consciously, others unconsciously. Some gestures have evolved into a social sign language understood by the majority of mankind. Others are related to rituals of religion, politics, business, and the military, and still others are associated with a particular nationality or with personal mannerisms and habits.

For purposes of analysis, there are three kinds of gestures: those that are functional, those that are expressive, and those that are both functional and expressive. Functional gestures include such actions as picking up a glass to drink from it, opening a door to enter into a room, and moving a chair closer to a table. The functional gesture may also be expressive, as in the example of a person who picks up the glass but pauses before drinking to admire the fine cut of the glass, the light reflected from it, or the aroma of the liquid in it.

Other typical functional gestures involve enumeration, pointing, and description. They are functional without the use of a prop. For example, a person may count on his fingers, hand and arm, head, or mouth. He may point to an object or another person, or in some direction. In describing something the person may indicate the height, width, distance, weight, size, or shape of something with his hands or indicate that something is rough, thin, thick, slow, or fast.

Other functional gestures signal someone to do something: for example, to come closer, go away, stand up, or sit down. Such gestures

61

may be expressive as well as functional. For example, shaking a fist at someone in anger may be a signal to stay away.

Some gestures have become symbols or signs whose meaning is immediately recognizable in almost every culture. Typical among such gestures are: shaking the head up and down or sideways to denote yes or no; shrugging the shoulders or raising the lower arms with the palms up to indicate ignorance or helplessness in response to a question; pointing and shaking the index finger at someone as a means of scolding or warning him; shaking a fist as a threatening gesture; raising the lower arm with the palm forward and one's head turned away to suggest refusal; performing the same gesture with one's head forward as a command to stop; beckoning someone with the index finger; indicating direction by extending the arm up, down, or to the side; and tapping a finger to the head to suggest either that one has an idea or that someone else should think about something.

Gestures of greeting or farewell can reveal social information and also attitudes between the people involved. Such gestures may denote, for example, social or military status, and coldness or affection. Greetings may take the form of a smile, a simple lift of the lower arm, a customary handshake, a "secret society" handshake, a military salute, an embrace with or without a kiss, a simple nod of the head, or a formal bow. Gestures of farewell can also take these forms; the most common, of course, is a wave of the hand.

Other gestures have become customary through common ritual, often religious, patriotic, or political. For example, in prayer there are the gestures of lowering and raising the head, raising the hands clasped or with the palms together, kneeling, and making the sign of the cross.

Among patriotic gestures in various countries are standing and placing the right hand over the heart for the national anthem, saluting the flag when one is dressed in a military uniform, and raising the right hand with the palm forward when one is being sworn into a political office or is in court.

Politically the raised fist has been associated with the Romans, Fascism, Communism, the Black Power movement, and Women's Liberation. The first two fingers spread apart has been used as a symbol for victory and for peace. To make the sign of the cross and to bless the faithful, the first two fingers are held together.

Some common gestures have their origin in sensory experiences. Disgust, derision, or contempt can be expressed by spitting, simulating regurgitation, sticking out the tongue, or making the "raspberry." To get someone to speak louder, one can cup the head behind his own ear or around the mouth. Lifting the head, widening the nostrils, and inhaling

62

can suggest that an odor smells good. Rubbing the belly and licking the lips can suggest that something looks good enough to eat. Rubbing the hands or fingers together can suggest the desire to get one's hands on something.

Some gestures can become personal mannerisms, as with the woman who continually pulls down on the sides of her dress or pats her hairdo, and the man who lifts the legs of his trousers before sitting or continually adjusts his tie. Among other gestures suggesting nervousness are stroking the chin, rubbing the index finger along the side of the nose, "wiping" the nose between the index finger and thumb, and pulling on the ear lobe.

In *A Psychology of Gesture*, Charlotte Wolff states that many of our gestures originate in periods of emotional development during childhood. She mentions a number of emotionally motivated gestures that can be useful to the performer in developing character movement patterns. Among them are:

1. Putting the hand or fingers in the mouth; biting the hand or fingers; biting the nails; wringing or interlocking the hands; clasping the hands or pressing one hand on the other; tapping the fingers; stroking or patting the face, the hair, or the nose; pulling on the hair; plucking the eyebrows; and fiddling with one's clothing.
2. Hiding the thumb in the palm; opening and closing the fists; stretching the hands; and cupping the hands to the head as if hiding.
3. Making mouth and tongue gestures; having facial tics; rolling the eyes; and making gestures of disgust regarding taste or smell.
4. Making gestures that are primarily away from one's own body; making an excessive number of unnecessary movements (which she refers to as mannerisms); demonstrating an almost total *lack* of movement; making emphatic or non-emphatic (passive) gestures; repeating the same gesture; and imitating the gestures of others who are present.

When observing others for gestures, watch for all the small signs that indicate how they actually feel and what they are privately thinking. Where are they looking? Are their arms and legs folded? Are their movements free or restricted and nervous? Are they tapping their feet or fingers? Do their facial gestures support what they are saying? When they talk, are their hands and arms in motion or still? Whenever possible, watch documentaries, films, newscasts, talk shows, and social situations where you can observe the gestures that people make on casual, formal, and ceremonial occasions, in their daily living, or when they think they are not being observed.

WALKING

Man ordinarily walks in an upright position, with his arms swinging in natural opposition to the movement of his legs.

The walk of each person is unique. People don't usually think about how they are walking and they rarely analyze why a particular walk is so characteristic of another person or his mental attitude. The way one walks can be affected by mental attitude, injury, illness, disease, age, and body build. The way in which one steps onto the foot; the use of the ankle, knee, hip, shoulder, and elbow joints; the carriage of the individual parts of the trunk; and the use of space and time all contribute to a characteristic walk.

In addition, the kind of clothing and shoes worn can alter the walk. For example, people walk differently when wearing high heels, low heels, boots, sandals, tennis shoes, no shoes, short dresses, long dresses, tight clothing, loose clothing, casual clothes, and dress clothes. A woman wearing her hair up, a hat, or long earrings, or a man in a top hat will generally carry the head high. To sense the changes try walking while balancing a book on your head or while wearing different kinds of shoes or clothing.

There are many different kinds of walks, but a few of the basic ones are those in which a person does the following:

1. Shuffles the feet.
2. Drags the toes or heels.
3. Steps on the whole foot, the ball of the foot, or the heel.
4. Bounces with each new step.
5. Keeps the center of gravity low, with the knees bent.
6. Keeps the center of gravity high and walks very smoothly.
7. Keeps the knees relatively locked for each new step.
8. Swings the leg well forward or backward with each step.
9. Walks with the legs far apart or close together.
10. Walks with short or long strides.
11. Swings the arms freely from the shoulder socket.
12. Swings only the lower arm with each step.
13. Keeps the arms immobile.
14. Walks with the legs turned inward or outward from the hip socket or knee joint.
15. Carries the trunk lifted high or slumped.

64

16. Pushes forward through space as if walking uphill.
17. Leans backward in space as if going downhill.
18. Walks at various speeds with personal rhythms.

Below are exercises for various ways of walking. They can help you to build characteristic movement patterns that are unlike your own.

THE NORMAL WALK

First it is useful to establish a norm for a basic walk. The norm can be considered the erect but not strained lift of the body in line with the center of gravity. The legs swing freely from the hip joints, with an easy bending at the knee and ankle joints. The heel of the leading foot reaches the floor first. The whole foot smoothly rolls down onto the floor, so that the opposite leg is freed to swing forward from the hip socket in the next step. As the free leg prepares to take the new step, the weight of the supporting leg is shifted to the ball of the foot to help propel the body forward. There is little vertical rise and fall of the body. The arms swing easily in opposition to the swinging of the legs. The movements are smooth, rhythmical, and relaxed. A smooth but slightly more stylized walk results when the ball of the foot reaches the floor first, followed by the lowering of the heel and the natural push off to the next step through the ball of the foot.

CHARACTER WALKS

The following exercises emphasize particular uses of the body that alter the normal walk in some way. When doing the exercises, emphasize first the action of the specific body parts under consideration while moving the rest of the body as naturally as possible. If space allows, walk in a straight line. Be aware of the emotional effect or attitude that each walk suggests. Decide which walks more effectively characterize certain kinds of people and age groups. Later practice combining various walks and building up characterizations from them. When you are preparing for a specific role, ask yourself how your character would use the various parts of his body in moving. The exercises will help you to determine the possibilities.

EXERCISE 69 *Use of the Foot*

a. Walk so that the heel strikes the floor first and then rolls down so that the whole foot is on the floor.

65

b. Walk so that the ball of the foot strikes the floor first and then lower the heel so that the whole foot is on the floor.

c. Walk so that the outside of the foot is placed on the floor first.

d. Walk so that the inside of the foot is placed on the floor first.

e. Swing the leg forward, kicking the floor with the heel before stepping onto the foot.

f. Swing the leg forward, kicking the floor with the toe before stepping onto the foot. (The walks in exercises 69e and 69f are sometimes done by children.)

g. Slide the whole foot forward along the ground before taking weight on the whole foot.

h. Swing the leg forward and step directly onto the whole foot. (Exercises 69g and 69h are often associated with the aged, whose joints have lost flexibility, and with those who have suffered from stroke or Parkinson's Disease, in which the tendency of the body to fall forward must be overcome.)

i. Walk on the balls of the feet, taking little if any weight on the heels.

j. Walk on the outside or inside of the foot, taking little if any weight on the other side of the foot.

k. Walk with the feet turned out, turned in, and then parallel. For the safety of the knees, the rotation should be taken from the hip sockets. In general the feet of the very young tend to turn inward, the feet of the old turn outward, and those of the late-youth to middle-aged group are somewhat parallel.

EXERCISE 70 *Use of the Knee Joint or Hip Joint*

a. Walk with a natural swing of the leg from the hip joint and with a natural bend of the lower leg. Try to keep the vertical rise and fall of the body to a minimum.

b. Walk with a minimum of action in the hip joint. Let the forward movement come from the action of the lower leg only. This walk is associated with the aged, who, with weight centered forward, step onto the whole foot.

c. Walk keeping the knees slightly bent. This walk can be used to characterize the aged, whose knee joints generally become stiff. It can also be used in a playful action such as sneaking up on someone, or for comic effect, as in the "Groucho Marx walk."

66

d. Walk with exaggerated bending and straightening of the knees so that the body has an effect of springing up and down, either smoothly or percussively.

EXERCISE 71 *Displacement of a Part of the Body*

a. Walk with the upper part of the pelvis tilted backward. This is sometimes called "tucking the hips under."

b. Walk with the upper part of the pelvis tilted forward. This causes a rounding in the lower back, often referred to as a "sway back."

c. Walk with the pelvis twisting from side to side with each step.

d. Walk with the pelvis shifting sideward toward the foot that is taking the weight.

e. Walk with the chest lifted and "puffed up."

f. Walk with the chest collapsed and rounded.

g. Walk with the torso bent forward from the waist. Bend forward to various levels.

h. Walk with torso bent backward from the waist.

i. Walk with the torso bending sideward at the waist toward the foot that is taking the weight.

j. Walk with either the shoulders, the torso from the waist up, or the whole torso including the pelvis twisting in opposition to the foot that is leading.

k. Walk with the chin dropped forward or lifted up.

l. Walk with the head rocking from side to side with each step. At first walk with the head bending toward the foot that is taking the weight. Later bend the head toward the opposite foot.

In general the very old tend to walk with the torso and head bent forward, the chest hollow, and the shoulders rounded. The walk of youth or the middle aged is affected by their habitual postural patterns and the physical activities in which they generally engage.

EXERCISE 72 *Distance Between the Legs*

a. Walk with the feet close together.

b. Walk with the feet a comfortable distance apart.

c. Walk with the feet widely separated.

d. Walk with the feet very wide apart.

Exercise 72a can also be done by placing one foot in a direct line in front of the other, as if walking on a tightrope or balance beam. The old tend to walk as in exercise 72a or 72d with small steps. People in the late teens through middle age tend to walk as in exercise 72b. Beyond this, however, distance between the legs varies with the individual, his habits, and his body build. For example, an obese person of any age tends to walk with his legs wider apart.

EXERCISE 73 *Length of the Stride*

a. Walk with small steps. Such a walk can be characteristic of the aged, the ill, or the tired. It may be used to sneak up on someone or to walk quietly. A woman wearing a tight full length dress would be required to walk this way. High heels worn by women or men in a period play tend to limit the length of the stride.

b. Walk with a natural comfortable stride.

c. Walk with a very long stride.

EXERCISE 74 *Use of the Arms While Walking*

a. Walk with a natural swing of the arms in opposition to the legs.

b. Walk with the same arm and leg moving forward simultaneously.

c. Walk with no arm movements (the arms held in place at the sides of the body).

d. Walk swinging the arms in close to the body, swinging the arms at various levels sideways away from the body, and swinging the arms across the front or back side of the body, in opposition to the forward steps.

e. As an exercise in simple arm-leg coordination, walk forward lifting each arm alternately to some level at the side. Try the exercise lifting the arm that is on the same side as the leg that is stepping out, and then try the exercise lifting the opposite arm.

Generally the younger a person is, the freer the movement in the joints will be. As a person ages the joints tend to stiffen. A young person will tend to swing the arms freely from the shoulder sockets. An older person tends to move the arms from the elbow and makes very little movement in the shoulder joint. The whole arm, or at least the upper arm, is usually held close to the body.

EXERCISE 75 *Speed of Steps and Dynamic Tensions*

Changes in the speed of the steps and dynamic tensions of the body alter the freedom of movement.

a. Walk in varying speeds and rhythms frcm slow to very fast.

b. Have a friend clap out various rhythms or play music with strong rhythms. Walk to them. Try to match the carriage of the body to the mood of the music. In walking to the music, change your steps: step on each beat, then on alternate beats, on the first or last beat of the measure, and on the offbeats.

c. Walk with the body very tense and stiff.

d. Walk with the body as relaxed as possible, almost as if the body did not have the support of the muscles.

e. Walk as if you are dizzy and fighting loss of control. This is a difficult physical problem that must be mastered if you are to play someone who is drunk. Often people playing drunks stagger violently around the area. Exactly the opposite approach is needed. The drunk is fighting *for* stability. If the problem is approached from this viewpoint, the appropriate balance between stability and loss of control can be found.

f. Walk with the knees slightly relaxed and bent and then, while walking, alternate very rapidly between tensing and relaxing the muscles of the thigh. If this is performed correctly, an unsteady shaking quality will result. If it is difficult to get the vibratory movement started, practice it first in place with the weight on both feet.

This vibratory control is valuable in walking or shaking with laughter or terror. It requires concentration and special coordination between relaxation and controlled tension.

EXERCISE 76. *Expression of Emotion While Walking*

Emotions and the state of the mind can alter the way a person walks. To suggest characters under the influence of specific emotions, you may wish to use some of the methods of walking described above. Keep in mind that the emotion must be internally motivated if the physical expression is to be convincing.

a. Walk with joy or happiness. Walk with the body erect, head lifted high, a bouyant lightness in the body, easy movement in the joints of the legs and feet, and free strides forward with arms swinging freely.

b. Walk with a sense of depression. Walk slowly with small steps, as if the body were weighted down. Bend the head and torso forward.

c. Walk expressing some degree of anger. There will be some variation depending on how fully the character is controlling his anger. In the case of extreme control, the body may be tense with the head slightly forward, as if the character were straining forward for an attack. The steps can be

69

slow and menacing, with an intense focus on the person or object causing the anger. If the character is abandoned in his anger, there may be violently jerky motions of the torso, erratic movements of the arms and head, and clenching of the fists and teeth.

d. Walk expressing fear. Fear can be characterized by the vibratory motion suggested above and by a feeling that the legs will not support one. Steps of this kind are slow and unsteady. Fear can also be characterized by cowering. The body may be bent over, with the arms lifted in self-protection and the head tucked down and turned either toward or away from the feared object. The direction of the steps may be sideward or backward. You may even want to break into a run away from the feared object. Fear which takes the form of timidity or shyness varies with the individual, but may be expressed by lowering the head or eyes and some rounding of the shoulders. The legs, or the lower legs only, may be turned in somewhat. Steps may be slow, small, and somewhat tense, with restricted arm and body movements, as if the character were trying to make himself small and insignificant in order not to be seen.

e. Walk in a manner expressive of pride or contempt. Walking with pride usually suggests an erect posture. The steps may be at any speed. If the idea is to express movement of the court or royalty, the steps may be slow and stately. The arm and body movements should be at a minimum. Easy confidence and informality may suggest a somewhat freer expression. The steps may be wide, graceful, and relatively free. From pride it is a short step to contempt, which may be expressed by lifting the chin and lowering the eyes to look down on those around you. The head may turn slowly from side to side.

EXERCISE 77 *Walking with a Change of Characterization*

In order to develop a kinesthetic sense of moving as a specific character, it is helpful to walk in a way that is expressive of that character. Changing the expressive walk in the following exercises should clarify certain distinctions in ways of moving that can alter the physical or psychological characterization. This could be helpful if, for example, you were required to be a young man in the first act of a play, a middle-aged man in act two, and an old man in act three. In doing the exercises walk from one end of the room to the other. Start the walk suggesting one characterization with your body carriage and then, as you continue to walk, alter characterization smoothly, so that by the end of the walk it has changed completely.

a. Start as a shy person and finish as a very confident person.

b. Start walking with bravado and finish crushed with defeat.

70

c. Start walking as a child and finish as an extremely old person.

d. Start walking as if your body were a balloon being inflated and finish as if your body were a balloon deflating.

e. Start walking as if your body were lighter than air and finish as if your body were as heavy as iron.

f. Start walking very rapidly and finish walking in slow action.

g. Start walking as a tall, thin person and finish walking as a short, stocky person.

h. Start walking as a person with elegance and grace and finish walking as a person who is bored and indifferent.

i. Start walking as a prancing show horse and finish walking as a lap dog.

j. If you are a woman, start walking as a young girl who is embarrassed by her new status as a woman and brings her shoulders forward to minimize her developing breasts. Finish as a mature woman confident of her womanhood. If you are a man, start walking as a cocky, strutting adolescent and finish as a mature man confident of his manhood.

k. Start walking as a person of royal blood and finish walking as a beggar.

l. Start walking as a rock star and end up walking as an ulcered business executive.

GENERAL WALKING AND GESTURING CONSIDERATIONS

Even a natural walk on stage has to be planned and rehearsed, to be done eventually without thinking, so that you are free to concentrate on other aspects of your performance. A few items for thought and practice in conjunction with walking, gesturing, and posing are presented here for consideration. You should always know which foot you will step out with first. If you are going to step out on a certain foot, the weight should at first be between the feet or on the other foot. You should practice taking the exact number of steps required to reach a new destination, so that there is not an awkward shifting of the feet before settling into a new pose or stepping onto stairs or some other object. The steps and the final pose should flow naturally into one another. When you are walking from one place to another, your balance will be more stable if you focus your eyes on someone or some object. If you are required to make a turn before walking, it is usually preferable to make the shortest turn to face the new direction. This may be a pivot turn on one foot, freeing the other foot to step out, or a pivot on both feet to face the new direction before stepping out.

71

Planning ahead can help. If you know that you must walk at a certain moment in the script or score, you can turn into a preparatory position one or two lines or beats ahead, so that you are ready to walk on cue. If the walk is to be interrupted, plan exactly the word or music cue you will step out on and the number of steps you will take before stopping.

The same planning should be done for gesturing, changing positions, and shifting weight. Certain moments in dialogue lend themselves to movement more than others. If the director leaves it up to you, decide when movement will give greater support or emphasis to the thought. As an example, suppose that the character says, "It is my opinion that. . . ." It may be more effective during that phrase to remain still, suggesting a moment of reflection or a gathering of thoughts before moving into actions that will support your opinion. On the other hand, if the character is somewhat indecisive, he may move about during the opening phrase, suggesting some confusion, and then stand firm when he expresses his convictions.

It is sometimes awkward for performers to walk up or down stage stairs, especially if, for aesthetic or practical reasons, the set designer has made the stairs too wide or too narrow for the customary stride of one step to one stair. It can be particularly awkward-looking if the performer must step on to the stair with one foot, bring the feet together, and then step onto the next level, unintentionally suggesting the gait of a child or aged person. There is really no rule for avoiding such problems, except to be aware of them. Work to make your movement look natural and generally unobstrusive. For example, you may be able to pause briefly every so many steps, or you may approach the steps somewhat sideways, so that two steps to each stair can comfortably be made. Ideally when you are walking up or down stairs the body should remain erect, the head lifted, and the eyes looking forward. The ball of the foot should be placed near the edge of the step.

EXERCISE 78 *Poise and Control in Walking: Blocking Designs (Figure 3-9)*
Practice is important to develop poise and control in walking and to familiarize yourself with the blocking designs developed by you or the director. Some exercises are presented below that will give you practice in walking, turning, and observing various stage directions. You should be familiar with various stage terms. *Downstage* means toward the audience. *Upstage* means away from the audience. *Stage left* is on the performer's left, and *stage right* on the performer's right when he is facing downstage. A *right turn* means to turn the right side of the body upstage (clockwise). A *left turn* means to turn counterclockwise. In order to help keep your balance when turning, focus your eyes on some object in the direction

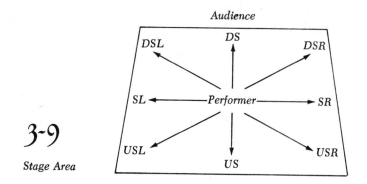

Audience

DSL · DS · DSR

SL ← —Performer— → SR

USL · US · USR

3-9

Stage Area

toward which you are turning. After performing each exercise below, repeat it starting on the other foot and turning to the opposite side. It is important that you become coordinated in your movements both left and right. Take one step for each count. For easier balance, you may at first extend your arms to the side at shoulder level.

In the following exercises a one-quarter turn (one quarter of a circle) equals a 90° turn (either left or right), a one-half turn equals a 180° turn, a three-quarter turn equals a 270° turn, and a full turn equals a 360° turn. The full turn brings you back to where you started.

3-10

Quarter Turn

a. Quarter Turns (Figure 3-10) Starting with the right foot, take four steps downstage, pivoting a quarter-turn (90°) to the right on the left foot on the fourth step. Repeat the sequence three more times. The floor pattern is a square. Repeat the sequence making a pivot turn on the third count. For some people the turn on the fourth count will be easier and feel more natural. (A pivot turn is made by placing the weight on the ball of one foot and turning on the foot so that you are facing in a new direction.)

b. One-Half Turns Starting with the right foot, take four steps downstage and make a full one-half turn (180°) to the right on the fourth step, so that you are now facing in the opposite direction (upstage). Then take four more steps, making another one-half turn to the right, to face downstage again. The floor pattern is a retracing of a straight line. Repeat the exercise, turning on the third step. For many people turning on the fourth step will be easier.

c. Intermittent Full Turns A full turn of the kind described here is not ordinarily required onstage; it is presented here as a coordination and balance problem. Starting with the right foot, take four steps downstage, making a full pivot turn (360°) to the right on the left foot on the fourth count. The right foot is lifted off the floor as you turn. End the turn by facing downstage. Repeat the sequence several times. The floor pattern

73

will be a straight line with a series of circles interspersed. Repeat the exercise, turning on the third count (right foot). For some people the third-count turn may be easier than the fourth. When making full turns, keep your eyes focused on some object in front of you as long as possible, and then when necessary quickly bring the head around and focus on the same object.

d. One-Half Turns Traveling in a Straight Line Another balance problem is to make consecutive half-turns while traveling across the floor. Although the effect when done quickly appears to be a full turn, the action really consists of two half-turns. Starting on the right foot, step forward and then make a one-half turn pivot to the right (180°), followed by a step back on the left foot, concluding the second half-turn with a pivot turn to the right on the left foot. Continue the one-half turn actions a number of times. Always step in the same stage direction so that you are moving in a straight line across the stage. Bending the knees will probably give you more stability at first. Later try it with the knees straight.

e. Stationary Full Turns Standing in one place with the knees bent or straight (whichever is easier), take four steps in place, making a full pivot turn to the right on the fourth count. Then repeat the step sequence and turn to the left on the fourth count. The floor pattern is a stationary turn on the vertical axis of your body.

3-11

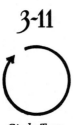

Circle Turn

f. Circle Turns (Figure 3-11) Facing downstage and starting on the right foot, take eight steps turning to the right: on each step turn one-eighth of a turn (45°), to face downstage right, then stage right, upstage right, upstage, upstage left, stage left, downstage left, and finally downstage. The floor pattern is a full circle.

3-12

Spiral Turn

g. Spiral Turn, Inward and Outward (Figure 3-12) Start on the right foot, take eight steps to walk in a large circle to the right, take another four steps making a smaller circle, take two steps making another even smaller circle, and conclude the spiral pattern turning in one spot. Reverse the action, turning in place and then making a small, a medium, and a large circle in order to arrive back at the beginning position.

3-13

Wave Pattern

h. Wave Patterns, Facing Stage Right (Figures 3-13) Starting on the right foot and keeping a straight line as your focus, walk stage right eight counts in one-half circle toward the audience, followed by a four-count

74

walk in a one-half circle away from the audience, followed by a two-count walk in a one-half circle toward the audience and a two-count walk in a one-half circle away from the audience, followed by a consecutive single-count pivot turn toward the audience, away, and toward, and away. The floor pattern is a wavy line in which the wave becomes progressively smaller.

i. Figure Eights (Figure 3-14)   Facing stage right, start on the right foot and walk stage right for four counts, making a one-half circle toward the audience, and then smoothly make a one-half circle upstage away from the audience, arriving at the beginning point (making a circle). Continue the walk sequence, making a one-half circle toward the audience, followed by a turn to the left, making a one-half circle away from the audience, and arriving at the beginning point (the second circle). The floor pattern traced by the walk should be a figure eight.

j. Triangular Floor Pattern (Figure 3-15)   Facing stage left and stepping out on the right foot, take four straight-line steps stage left. Pivot on the fourth count one-quarter of a turn to the right and continue the four-count step pattern in a straight line forward. Pivot on the fourth count three-eighths of a turn (135°) to the right and continue the four-count step pattern in a straight line back to the beginning position. The floor pattern should be triangular.

k. Walking Backwards   The purpose of this exercise is to develop your sense of the space around you and your peripheral vision. Walk backward toward some object such as a couch or wall. Without extending your arms backwards or looking sidewards or backwards, trust your sense of spatial orientation and your peripheral vision to tell you when to stop walking so as to come very close to but not make contact with the object behind you.

l. Side Step-Close   Facing downstage, step to the right side on the right foot. Bring the left foot to the right. End with weight on both feet. Repeat several times. The floor pattern traveled will be a straight line to stage right.

m. Grapevine Side Step   Facing forward, step to the right side on the right foot, and then step with the left foot, crossing over in front of the right foot. Repeat the side step to the right, and alternate the crossing of the left foot each time front and back. The same grapevine step while facing forward can be done traveling either forward or backward, alternately crossing each foot over the other as if walking on a tightrope. The floor pattern will be a straight line sideward, when stepping sideways.

n. Box Step   Facing forward, step forward on the right foot, then close

3-14

Figure Eight

3-15

Triangular Floor Pattern

the left foot to the right foot. End with the weight on both feet. Next, step to the left side on the left foot and close the right foot to the left foot. Repeat the action, stepping back on the left. Close the right foot to the left foot. Step right on the right foot, closing left foot to right foot. The floor pattern is a square.

o. Change of Level Facing forward, step forward in sequences of four counts each with the arms lifted to the side at shoulder level. For the first four steps, lower the body to a deep knee bend and for the next four steps to a medium knee bend. The next four steps are done with straight knees. For the next four counts rise on the balls of the feet with the knees bent, and for the final four steps rise on the balls of the feet with straight knees and with the arms stretched over the head. Keep the torso erect throughout and try to keep the body stretched vertically upward. Try the same sequence walking backward or sideward.

For variety, combine some of the exercises suggested above, change the number of counts walked in each direction, change the floor pattern, vary the tempo of the walk, vary the turning pattern, vary the emotional or dynamic content, and use the arms in different ways.

For example, while angry take three steps forward toward a large dog who is tearing up your garden, stop, do a deep knee bend, pick up a large stick, rise, and take four slow steps backward (in response to the snarling dog, who is slowly moving toward you), make a quick half-pivot turn, slowly run for eight steps in a straight line, and then run at full speed in a circle as the dog snaps at your heels. The first time play the scene realistically and the second time for comic effect.

*By his controlled use of gesture
the actor transforms the floor into a sea,
a table into a confessional, a piece of iron
into an animate partner . . . .*

JERZY GROTOWSKI

4 Working with a Group

WARMING-UP EXERCISES

The average person feels somewhat awkward and inhibited about moving. With experience these feelings about moving can be overcome, like other problems of the performer's craft. The following exercises can loosen up your body and develop strength, flexibility, and control, all of which can help you feel more comfortable when moving. These exercises can also be used to warm up or prepare your body before rehearsing or performing, just as singers and actors warm up their voices with vocal exercises. Warming up the body and releasing undesired tensions can help to keep the body more "alive" and make a livelier performance.

The following exercises can be done alone or with a group. Doing these or similar kinds of exercises to music or counts with a group can help establish the disciplines of following the musical beat and working in unison with others. If you are working with a group, someone can act as leader and decide how many of each action or combination of actions should be done, how fast, and whether they should be done to a specific number of counts or whether music (on records) can be used. It is better to do the actions without excess tension or undue strain in order to free your movements. Later you may want to use these exercises as a basis for more demanding exercise patterns.

FACE EXERCISES

EXERCISE 79

Open the mouth and eyes as widely as possible, and then close them as firmly as possible. Repeat a number of times until the face tingles.

EXERCISE 80

Slowly and silently count from one to ten, exaggerating the opening of the mouth and the use of the tongue and lips.

EXERCISE 81

Silently speak the scale (do, re, me, fa, so, la, ti, do) several times with exaggerated facial movements. Then silently speak the letters of the alphabet, also with exaggerated facial movements.

STANDING EXERCISES

Do the following exercises with the feet apart to the side and then with the feet apart to the front and back, unless indicated differently in the directions below.

Shaking

EXERCISE 82

With as little tension as possible vigorously shake the hands as if shaking water off them. Repeat the action with the whole arm, with each leg, with the head, and finally with the whole body.

Swinging

EXERCISE 83 *Head Swing*

Relax the head, letting it fall forward so that the chin rests on the chest.

Gently roll the head from side to side across the chest. Increase the range of the head swing until the face points to the side of the room on each swing. Increase the range more so that the face is directed upward to the ceiling on each swing.

EXERCISE 84 *Arm Swing*

Starting with the arms hanging loosely at the sides of the body, swing one or both arms forward and backward like pendulums. Increase the range of the swing until the arms are above the head. Repeat the exercise, swinging the arms across the body in front and then in back.

EXERCISE 85 *Leg Swing*

a. Stand erect with the weight on both feet and the toes directed in a line forward of the body. The arms can be extended to the side at shoulder level to help your balance. Shift the weight to one foot, freeing the other leg. Loosely swing the free leg forward and backward like a pendulum. Swing the leg only as high as is comfortable so there is no strain. Stop and repeat the leg swing, this time swinging the free leg sideways across the body and then away from the body. The purpose of the leg swing is to free the leg and hip joints of unnecessary tension.

b. Swing the lower part of the leg. The swing can be done more easily with the lower leg while seated on something so that the feet are clear of the floor. With no inward or outward rotation of the leg, swing the lower leg backward and forward. Repeat the action, alternately rotating the upper leg inward from the hip socket as the lower leg swings forward and rotating the upper leg outward as the lower leg makes the backward swing.

EXERCISE 86 *Torso Swing*

a. Bend over at the waist with the arms and head hanging loosely toward the floor. Swing the whole torso from side to side like a pendulum. Increase the range of the swing, making sure that the torso comes up to an erect vertical position on each swing.

b. Repeat the action, swinging the arms in the same direction as the torso swing.

c. Repeat the action, swinging the arms in the direction opposite to the torso swing.

d. Starting with the body erect, the feet separated to the side, the arms extended above the head, swing the torso forward and down toward the knees. Return to the upright position swinging forward and up.

79

Circling

EXERCISE 87 *Head Circle*

Relax the head so that it is forward, with the chin on the chest. Roll the head toward the right until the ear is centered over the right shoulder, roll the head toward the back so that the face is directed toward the ceiling, roll the head toward the left so the ear is centered over the left shoulder, and roll the head forward until the chin is resting on the chest again. The head has now made a complete circle to the right. Repeat the action a number of times to the right, moving smoothly and continuously through each position, and then repeat the head circle to the left.

EXERCISE 88 *Torso Circle*

a. In a pattern similar to that of the head roll, move the torso in a circular path around your base of support, the feet. Start with the hands on the hips. Lean slightly forward at the waist, with the torso lengthened in a straight line. Move the torso in a circular path to the right side, to the back, to the left side, and to the front. Repeat the torso circle a number of times to the right and then to the left. Repeat the torso circle with the arms extended to the side, forward of the chest, and over the head.

b. Increase the range of the circle by starting with the torso hanging loosely toward the floor and the arms stretched toward the floor (figure 4-1). Swing the torso to the right side, bending sideward at the waist with the arms stretched above the head. Move the torso to the back and to the left side, and then finish with the torso swinging down to the forward beginning position. Repeat the movement smoothly a number of times to the right and then to the left.

4-1

Exercise 88b:
Torso Circle

80

EXERCISE 89 *Arm Circle*

a. Starting with the hands hanging loosely at the sides of the body, swing one or both arms forward, up over the head, and down in back. Repeat the circular swing a number of times with the arms relaxed throughout. Repeat the swing to the back, up over the head, and down in front.

b. Repeat the swing pattern several times, moving the arms to the sides away from the body, up over the head, and down by crossing the arms in front of the body. Repeat, reversing the direction, swinging the arms across the body, over the head, and down to the sides away from the body. The crossing can also be done in back of the body in the side swings, although it will require some bending of the torso and arms.

EXERCISE 90 *Lower Arm Circle*

a. Start with the upper arms held against the side of the body and the lower arm bent forward to a right angle. As if you were drawing circles on the wall in front of you, move the lower arm in a circular pattern in one direction and then in the other direction. Try the same pattern (a sort of stirring action) using only the hand, then the lower arm, and then the whole arm.

b. Try the same actions as if you were drawing the circle on a wall at your side or with the arm straight in back of you. Try the action first with the hand, then the lower arm, and then the whole arm. Repeat the circle in both directions.

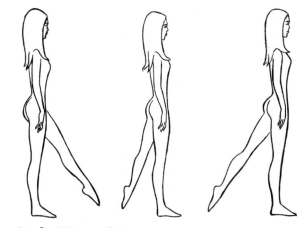

4-2

Exercise 91: Leg Circle

EXERCISE 91 *Leg Circle* *(Figure 4-2)*

a. Extend one leg forward of the body with the toes touching the floor. Lengthen the leg so the leg makes a straight line from the hip to the tip of the toes. Tracing a pattern of a half of a circle on the floor, move the

81

leg to the side, to the back, toward the supporting leg, and again to the forward position. Repeat several times and then reverse the direction.

b. Repeat the half-circle pattern with the foot lifted off the floor and the leg relaxed. Try to do the pattern with the same free action you used with the arm swings.

All of the above circling actions can also be done in figure-eight patterns, that is, tracing the number eight around the body.

Twisting

EXERCISE 92 *Head Twist*

a. Start with the face directed forward. Sharply turn the head to the right, forward, to the left, and forward. Repeat a number of times, making certain that the eyes focus on something directly forward of the face.

b. Repeat the head action, looking up, forward, down, and forward.

c. Combine the two patterns together. Look right, up, left, down. Repeat several times on both sides.

EXERCISE 93 *Torso Twist*

a. Rotate the shoulders and chest to the right and to the left a number of times in a continuous rotary action. Repeat the action in the torso from the waist up a number of times, and then with the pelvis by itself a number of times.

b. Turn the head to the right, followed by the torso from the waist up, followed by the pelvis. Repeat the same sequence to the left.

c. With knees slightly bent, twist the head and torso to the right at the same time that the pelvis is twisted to the left. Repeat a number of times, twisting the two halves of the body in opposition to one another and alternating the sides.

Swaying

EXERCISE 94 *Torso Rock*

a. Separating the legs forward and backward, and keeping the torso lengthened in a straight line, bend the torso forward and then backward in a continuous motion, as if seated in a rocking chair. Separating the legs to the side, repeat the same action, rocking from side to side.

b. Repeat the action with the whole body, using the feet. When rocking forward, lift the heels slightly from the floor, and when rocking backward lift the toes slightly from the floor. When rocking sideward, take the weight on the outside of the foot in the direction in which you are moving.

82

EXERCISE 95 *Step-Sways*

a. Starting with the body lengthened upward, alternately step from one foot to the other foot. Move the torso in a smooth swaying motion in the direction of each step. Step to the right side on the right foot, to the left side on the left foot, to the front on the right foot, and to the back on the left foot. Repeat a number of times. Repeat the entire exercise starting to the left on the left foot.

b. Repeat the pattern with the knees bent throughout and with the arms swinging in the direction of the step.

c. Vary the exercise by stepping only on the balls of the feet, only on the heels, and only on the whole foot.

Stretching

EXERCISE 96 *Stretch-Relaxation (Arms)*

With the fingers spread apart, stretch both arms sideward away from the body as if the body were being pulled apart into two halves. Maintain the stretch action for a count of six and then release all of the tension from the arms and let them drop to the sides of the body. Relax them for a count of six. Repeat several times, and then repeat the stretch-relaxation with the arms forward of the chest, backward, over the head, and toward the floor.

EXERCISE 97 *Stretch-Relaxation (Whole Body)*

Repeat the above stretch-relax pattern by reaching over the head for six counts, and then relaxing and falling toward the floor into a deep knee bend. Hold the relaxed position for six counts. Repeat.

4-3

Exercise 98: Leg Stretch

EXERCISE 98 *Leg Stretch* (*Figure 4-3*)

Stand with the feet apart, either to the sides or with one back and one forward. Bend the torso forward toward the floor. Stretch the arms toward the floor. Do a deep knee bend, keeping the heels on the floor, and place the hands on the floor for additional support. You should be in a "squat"

83

position. Look backward through the legs. Continue to look backward through the legs while straightening the legs. If your muscles are not sufficiently stretched you may have difficulty getting the legs completely straight. Only straighten the legs to the point where there is minimum discomfort. Repeat the bending and straightening action a number of times. You should feel most of the stretch at the back of the legs.

Bouncing

EXERCISE 99 *Shoulders*

Lift both shoulders up and hold, then let them drop down into place and relax for a moment. Repeat this action until the shoulder area feels somewhat relaxed. Then lift and drop the shoulders at a faster continuous rhythmical pace, somewhat like a bouncing ball. Keep the movement as free of tension as possible.

EXERCISE 100 *Arms*

Lift one or both arms to the side and hold. Then let them drop of their own weight down beside the body and relax for a moment. Do this a number of times until the arms are completely free of tension. Repeat the sequence at a faster rhythmical pace. The arms should feel as if they had neither bones nor muscles to them.

EXERCISE 101 *Whole Body*

Stand with the feet apart. With the toes straight forward, bend the knees slightly so that the knees are directly over the middle toes. Keeping the knees bent, rapidly and rhythmically bounce up and down. Keep the head, torso, and arms as relaxed as possible.

LYING DOWN EXERCISES

EXERCISE 102 *Stretching*

a. Lie on your back with your arms over your head and resting on the floor. Stretch the arms and at the same time stretch the legs in the other direction. Hold for a count of six. Relax for six counts and repeat the exercise several more times.

b. Stretch the legs and arms to the side as far as possible. Hold the position for six counts and repeat several times. In both exercises make certain that the lower back remains on the floor. On the stretch, think of lengthening the leg in one straight line from the hip to the tip of the toe.

84

4-4

Exercise 103:
Upper Body Lift

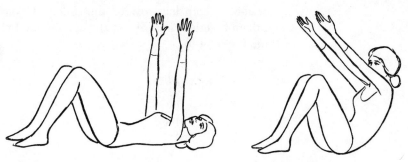

EXERCISE 103 *Upper Body Lift (Figure 4-4)*

Continuing to lie on the back, bend the legs so that the knees are together and facing toward the ceiling, the heels approximately ten inches from the buttocks, and the arms extended forward of the chest. Lift the head, then the shoulders, and then the upper and lower back from the floor until a tightening in the abdomen is felt. Hold for six counts and then in succession lower the upper back, lower back, shoulders, and head to the floor. Relax for six counts and repeat several times. If you are able to keep both feet on the floor, repeat the exercise with the legs stretched out rather than bent.

4-5

Exercise 104a: Pedaling

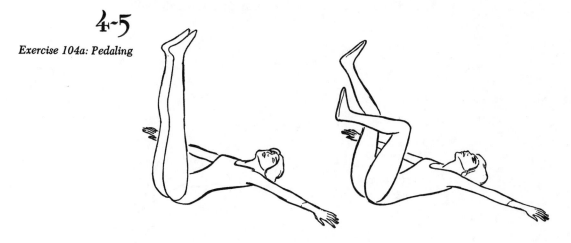

EXERCISE 104 *Leg Raises*

a. Lying on your back, raise both legs toward the ceiling (figure 4-5). Keep the lower back on the floor. As if you were riding a bicycle, alternately bend and straighten the two legs. As one bends the other is straightening. Repeat this a number of times as if you were pedaling a bicycle up a hill. Repeat the action faster with a minimum of tension.

b. Straighten both legs toward the ceiling. Open both legs to the side, bring them together, and repeat the opening and closing action a number of times (figure 4-6).

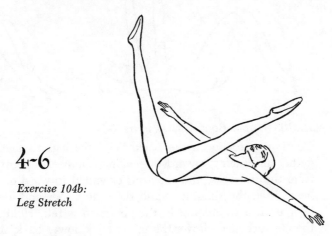

4-6

Exercise 104b:
Leg Stretch

c. Repeat the same kind of action forward and backward in a manner that is similar to the kicking action used in a basic swimming kick (figure 4-7). Keep the lower back on the floor throughout.

4-7

Exercise 104c:
Kicking

SITTING EXERCISES

EXERCISE 105 *Ankle and Leg Stretches*

    a. Sitting on the floor, straighten the back and lengthen the torso upward while the legs are lengthened in front of you in a straight line from

4-8

Exercise 105a:
Ankle Stretch

4-9

Exercise 105b:
Ankle and Leg Stretch

the hip to the tip of the toes (figure 4-8). Keeping the legs straight, bend the foot toward you at the ankle joints until the heels are off the floor and the calf muscles feel tightened. Hold the position for six counts and return the foot to the stretched position for a count of six. Repeat several times.

b. Repeat the bend at the ankle joint, at the same time bending the legs so that the knees and the tips of the toes are directed toward the ceiling (figure 4-9). Return to the stretched leg position. Repeat several times.

c. While bending the legs and ankle joints, bend the torso forward toward the knees and clasp the feet with the hands (figure 4-10). Return to the upright position as the legs stretch out. Repeat several times.

4-10

Exercise 105c:
Body Stretch

4-11

Exercise 106:
Long Body Stretch

EXERCISE 106 *Long Body Stretch (Figure 4-11)*

Sitting on the floor with the torso erect, the legs stretched in front, and the arms stretched over the head, bend the torso forward and clasp the feet if possible. Hold for six counts, return to the upright position, and hold for six counts. Repeat a number of times.

87

Sitting on the floor, lean back and support yourself on your hands or lower arms. Stretch both legs out in front of you. Lift one leg up, carry it to the side away from the body, lower it approximately one inch from the floor, bring it toward the other leg, and lower it to the floor (a circular action). Repeat the action in reverse: lift the leg one inch from the floor, carry it to the side away from the body, lift the leg higher, carry it toward the center, and lower it to the floor next to the other leg. Repeat the exercise several times with one leg and then the other.

4-12

Exercise 108: Leg Lifts

EXERCISE 108 *Leg Lifts (Figure 4-12)*

Sitting on the floor as you did in exercise 107 above, lift one extended leg off the floor sixteen times and then repeat with the other leg. The leg should be lifted high enough to feel an easy pull at the back of the thigh. The leg lifting action should also be done while lying on the front or the side of the body.

GROUP SENSITIVITY

In the civilized state man rarely uses his senses fully in communicating, as other animals do. We are taught that there are preferred ways of communicating with one another: primarily through sight and through the verbalization of ideas.

From an early age we learn not to explore our own or other people's bodies by touching. As a result it is sometimes psychologically difficult for people to touch or to be touched. We usually limit our physical contact to a quick handshake or occasionally an embrace, and rarely explore the texture of someone's skin or hair with our hands. Even in sexual contact touching is frequently limited to the erogenous zones only.

Only certain kinds of odors are considered socially acceptable. When we are told we smell good, others are usually referring to our cologne

88

and not to us. We try to remove or disguise our natural body odors by bathing regularly or wearing an artificial odor. In contrast, some primitive tribes smell one another behind the ear when they meet. We accept it as natural if a pet dog or cat smells or licks us, but would be startled or perhaps even outraged if another person did the same thing.

There may be persuasive arguments as to why we limit the use of our senses in daily communication. However, as performers we must recognize when cultural inhibitions are interfering with desirable interactions with our fellow performers. We have to learn to break down psychological barriers and accept at least recognition of our common biology.

Good ensemble work is difficult to achieve because people usually come together for one production and then separate. There is never time to develop fully the kind of performance interaction that a repertory group performing together in production after production can achieve. Good ensemble work requires that every member of the company be receptive to the strengths and weaknesses of every other performer and willing to work with the others for the common goal of the production.

To work successfully with others requires that your own senses be finely attuned to yourself, to your surroundings, and to the needs of the other performers. Some of the exercises used in group therapy and sensitivity training can be very helpful in developing your ability to work as individuals within a group.

The following exercises should be thought of as experiences. They cannot be done rightly or wrongly. Their only value is to develop your own awareness of yourself as a sensual human being who is willing to share this experience with others by both giving and receiving. Each group should have a leader who will silently direct the actions of the others to establish a group purpose. Do not talk during the experience; feel it. Later you may want to discuss how you felt. Although the exercises are for a group, some of them can be done by yourself to yourself for a similar purpose—self-awareness.

The experiences are of four kinds: touching, supporting, verbal embraces, and leading-following. Each person should take a turn being in the center of the group.

EXERCISE 109 *Touching*

a. The volunteer lies down on his stomach, closes his eyes, and completely relaxes. The others kneel around him. The group leader chooses some way of touching the volunteer. Among the choices are vigorously patting with the whole hand, gently slapping with the fingers, tapping

89

with the fingertips, and massaging with the whole hand. This should be done gently either to stimulate the nerves under the skin or to relax the body totally, as in a massage. It should never be done so hard that the person becomes tense. Explore the whole body. The exercise should be thought of as a pleasant experience for both people.

b. The volunteer should stand in the center of the group with his eyes closed. One by one each person in the group moves to the volunteer and touches him in some way that will evoke a pleasant tactile experience for the giver and the receiver.

c. With eyes open, the volunteer makes eye contact with each member of the group and then moves toward him in turn. The two should express in some way the kinds of feelings their coming together generates. It may be a pat on the back, a gentle embrace, a light kiss on the cheek or some other sincere demonstration of physical warmth to the other person. Whatever the reaction is it should be natural and spontaneous, but never negative.

d. The group should pair off. Facing the other with the eyes closed, each member should explore the other's face. Each should be aware of the sensation of being touched and touching and of the shape and arrangement of the parts of the face and the various rough and smooth textures.

e. A verbal variation on giving and receiving is to join hands in a circle around the volunteer, who is in the center of the circle with eyes closed. The group then whispers over and over the name of the person in the center and slowly moves in toward the person. The volunteer's reaction may be increased if he is seated and the group is standing. The first time through, whisper with a warm, loving expression. Later try abstract or literal sounds as well, such as humming, laughing, or animal sounds. After each experience discuss the reactions of the group and of the volunteer. Try to keep the experience pleasant for the receiver.

EXERCISE 110 *Supporting*

These are exercises in which a person is physically supported or moved from one person to another. Their purpose is to develop trust in the volunteer that the group will not drop him, so that he can relax his body during the exercise. The volunteer should allow his movements to be controlled by others.

a. At the start the volunteer stands in the center of the circle, with the group close to him. The group moves him back and forth or around the circle. The group should never let him lean over more than six inches, so that his balance is never completely lost. The volunteer should keep only enough tension in his body to remain standing. Passively he should allow himself to be moved by the group.

b. The volunteer stands between two people, who gently push him back and forth. Each should stand one foot from the volunteer. For the exercise to be effective, the volunteer must surrender his resistance, allowing the work to be done by the pushers.

c. The volunteer lies on his back on the floor with the group kneeling around him. The group then lifts the volunteer up over their heads. If the group members are secure in their support, the volunteer can be bounced gently up and down in that position. Start with a lightweight volunteer or strong lifters.

d. In an easier version of this, the volunteer lies on his back and the group breaks up into smaller subgroups. One or two people work together to lift the person's head, arms, legs, or some other part of his body. The volunteer should completely relax and let the lifters do the work.

EXERCISE 111 *Following and Leading*

a. In pairs facing one another, with the eyes open and fingertips touching those of the other person, move together to explore as much space around each other's body as is possible. Move in all directions: upward, downward, forward, backward, sideward, and diagonally. Move slowly. As much as possible, try to avoid forcing the movement by yourself. The idea is to move together as one person rather than two separate people.

b. The two members of the pair face each other. One person slowly moves his arms or whole body in some way through space. The other person, as a mirror image, performs exactly the same movements at exactly the same time.

c. Another exercise of this type is to lead someone through as many pleasant experiences as possible. The person being led closes his eyes. The leader holds his hands and leads him carefully from experience to experience, watching out for his safety and seeing that he does not trip or fall. A sensory experience was conducted by the Company Theatre in Steven Kyle Kent's James Joyce Memorial Liquid Theatre, where each member of the audience with eyes closed was led from one experience to another. Some of the experiences were a spray of perfume, a grape placed in the mouth, a hand bathed in cool water, water sprinkled on the face, blowing on the cheek, gentle kisses, and warm embraces. These kinds of experiences can be and should result in an awakening of senses that are too much neglected.

GROUP INTERACTION

The purposes of the following exercises are to develop a sense of rhythm and to develop the ability to respond quickly to gestures and sound.

EXERCISE 112

The group sits in a circle with joined hands. As group leader, one person squeezes the hand of the person next to him, who in turn squeezes the hand of the person next to him, and so on, sending the impulse around the circle and back to the leader. After this has been done a number of times, the leader then sends the impulse simultaneously in both directions around the circle. Ideally it should arrive back to the leader from both directions at once. It may take practice, but it can be done.

EXERCISE 113

The group remains in a circle. The group leader makes some kind of simple gesture, which is in turn repeated by each person around the circle. The group should try to keep the gesture exactly as it was when originally done. At first watch each person who is sending the gesture, but later watch only the person sending the gesture to you and the person to whom you are sending it. It is better at first to keep the gestures clear and well-defined, like a shrug of the shoulders, a tilt of the head, or a lift of the arm. Later you may want to join several gestures together to express a simple thought, such as "I don't know," "I am sleepy," or "I feel great."

EXERCISE 114

This is a variation on passing the gesture that requires more concentration on quickly memorizing and repeating movements. The group leader makes a gesture which is repeated by the person next to him, who in turn adds a gesture of his own. Each new receiver around the circle repeats each gesture that has been done previously and adds one of his own. Try to repeat the gestures exactly as they were done and with the same kind of expression.

EXERCISE 115

Pass an imaginary object around the circle. Each person should maintain the exact shape, size, and texture of the imaginary object and react appropriately to what he thinks it is. For example, it may be circular and heavy, square and big, tubular and long, or small and alive.

Another passing exercise combines sound and gesture. One person makes some kind of distinctive sound: loud, soft, quick, or perhaps drawn out. The person receiving the sound responds to it by executing a movement that has the same expressive quality as the sound. Then the receiver passes a new sound onto the next person, who in turn responds and passes another sound onto the next person, and so forth. Instead of sounds, words that evoke images or emotions may be used.

A real object, such as a ball, is passed around the circle. The ball should be received by one person in one count and passed onto the next person on the next count. *Before* receiving and *after* passing the ball the person claps with the group, who are clapping to keep the rhythm steady. The point is to keep the ball changing hands on every count. The rhythm of the clapping can be gradually speeded up to make the exercise increasingly difficult. Slightly more difficult is to have two balls passed around the circle in opposite directions.

Next do the same exercise but alternate the claps with isolated body movements or arm gestures. For example, clap on count one, lift the right arm over the head on count two, clap on count three, lift the left arm over the head on count four, clap on count five, extend the right arm forward of the chest on count six, clap on count seven, extend the left arm forward of the chest on count eight, and so forth. Keep the movements simple and keep repeating them in some kind of pattern. The ball in the meantime continues to be passed around the circle in rhythm with the claps and gestures.

The purpose of the following three exercises is to cause the group to move spontaneously together.

Sitting close together in a small circle, all members of the group place their right hands in the center so that all the hands are joined together. Each person then places his left hand on the shoulder of the person to his left. Without anyone dominating the movement individually, try as a group to move in place, kneel, stand up, or travel around the room. Move slowly, as if the group were a single person rather than a group of individuals. Try changing the group shape and size in as many different ways as possible. Try different kinds of actions, such as swaying, swinging, bouncing, stretching, and circling.

EXERCISE 120

This group exercise uses the mirror image concept. The entire group divides in half. The two groups form straight lines across the room and face one another. Each member joins hands or fingertips with the people next to him in his own group. One group moves *as a group,* and the other group tries to mirror its actions. The members of each group should be aware and sensitive to the actions of their own group as well as the group opposite them.

EXERCISE 121

The group joins hands in a circle. The group leader speaks a sound, a word, a phrase, or a sentence several times with the same identifiable dynamic or emotional quality. The group then moves as one in a way that resembles the vocal quality of the speaker. For example, if the speaker were to speak in a whisper with very pleasant sounds, the performers might move smoothly and softly. If the speaker were then to speak with percussive sounds, the performers might move with staccato actions. If the speaker sounded angry the group might move as if angry or agitated. The value of the experiment is to listen to vocal quality, react quickly, and express it in body movement as a group.

IMPROVISATION

Improvisation is valuable in a number of ways. One of its primary functions is to learn to think, to feel, and to react spontaneously without pre-planning. In improvisation a performer has to learn to think on his feet and respond to the unexpected as presented by another person or the group. Improvisation can tap unexplored emotional resources in developing a role. It has the potential to reveal to each person that which makes him unique as well as that which unites him with all mankind. Other benefits are those of using the body or voice in a way that might not occur if the actions originated from the conscious mind only. It can also free you from inhibitions and allow you to move for the sheer joy of self-expression regardless of whether the movement is appropriate or not.

Improvisation requires freedom, concentration, and a total commitment of energy in order to be successful. There are many different ways to improvise. Some improvisational exercises are presented below that can be explored by yourself or, preferably, with others. Other improvisational situations are suggested throughout the book. In order to get the most out of the improvisational exercises, note the following suggestions.

1. When involved in the improvisation, abandon yourself to the idea being expressed and block out your awareness of anyone who might be watching you.

2. Do not consider that there is a right or wrong way of doing the improvisation, but rather consider it as an exploration of one or more of the infinite possibilities of a given situation.

3. Do not plan out the details of what you are going to do, but rather let things happen spontaneously. Be willing to follow your impulses and feelings.

4. When you are working with another person, try to respond to his psyche, emotions, and physical movement. This response is not unlike what you do in everyday life. When someone is expressing an emotional reaction, you pick up either consciously or unconsciously his "vibrations" and react in some way.

5. If the improvisation is built around a theme, try to shape it so that it builds to a climax, with a beginning, a middle and an end.

6. In general, when beginning the improvisation do not immediately start moving and expressing yourself, but take a few moments or longer to block out what is going on around you and focus your concentration and energies on the improvisational situation.

7. Do not allow yourself to wonder how well you are acting out the improvisation or how well anyone else thinks you are acting out the improvisation. If you can free yourself of your own and others' value judgments and abandon yourself to the idea to be expressed, you will find that improvisation can be a very rewarding and an enjoyable experience.

The exercises that follow are based on word or phrase imagery. Following are lists of words or phrases grouped according to some general concept. By selecting a number of words or phrases from one or several categories and then as a group establishing each word or phrase idea in movement, the group can achieve a sense of ensemble playing. You can get a feeling of the overall physical shape or mood of the group and the way in which you as an individual help determine the end result. Selected words can also help determine a way of moving or developing a movement phrase for a characterization.

If you are working alone rather than with a group, select five or six words or phrases and write them down. Then move in some way that will suggest how you respond to the imagery of each word. Do not preplan your moves. Try to imagine yourself as representing the word rather than moving as a person. Make the movements broad and free. Think of the shapes your body is making in space. Try to move smoothly from one movement to the next without stopping the flow of your actions. Remember that the idea is to express yourself spontaneously through movement and not to give a rehearsed stage performance.

If you are working with other people, divide into groups. One group should watch and set up the improvisational situation for the other group. The working group should be given clearly defined goals to achieve together as a group. If necessary they can discuss briefly how they will work, but it is more challenging if no discussion takes place and they are forced to reach the goals by communicating silently. Specific exercises follow the word lists.

WORD IMAGERY FOR IMPROVISATION

Smell	*Taste*	*Touch*	*Sound*
acrid	acidic	blunt	blaring
foul	bitter	brittle	groan
fragrant	dry	bumpy	gurgle
rotten	pungent	dull	melodious
stale	rancid	greasy	sighing
sweet	spicy	moist	strident
vile	tart	sharp	twang

Shape	*Action*	*Gesture*	*Time*
clustered	creep	catch	adagio
contracted	fall	greet	briskly
expanded	roll	lift	extended
high	sneak	reach	hurried
low	surge	reject	leisurely
round	swirl	slap	measured
square	wiggle	throw	rushed

Characterization	*Attitude*	*Emotional*	*Weather*
arrogant	boorish	ardent	blizzard
confused	cunning	calloused	foggy
drab	cynical	caustic	sprinkles
languid	devious	eager	tempestous
lethargic	enthusiastic	furtive	tornado
lusty	jaded	frenzied	turbulent
robust	vicious	greedy	whirlwind

96

Animal Sound	Mood	Animal	Body
bellow	compliant	cow	fragile
grunt	contrite	elephant	gross
hiss	gloomy	fox	plump
screech	irritable	giraffe	stocky
snarl	lighthearted	rat	svelte
squeal	morose	snake	tiny
whine	pensive	weasel	wiry

Age	Non-Human	Phrases
callow	atomic blast	boom of jets
codger	boiling water	crack of the whip
coltish	gurgling brook	falling leaves
decrepit	erupting volcano	pattern of rain
dowager	evolution	roar of the surf
filly	rising sun	ticking clocks
hag	wind storm	weeping women

The following exercises suggest ways to proceed for each category of words and phrases.

EXERCISE 122 *Shape*

a. To develop a sense of group floor patterns, the group selects four shapes, such as the letter B, a triangle, a half-circle, and the numeral 4. The working group without any preplanning moves into each of these shapes. Figure 4-13 shows an example. They may establish each one indi-

4-13

Exercise 122:
Shape Exercise Example

vidually or break into smaller groups and do all four at the same time. They may do them while standing, kneeling, sitting, lying down, or in some other position. The more creative exploration, the better.

97

b. To develop a sense of group shape, the group selects four shapes, such as a sphere, a star, a rectangle, and a cube. The group tries to shape a solid mass rather than an outline. The group moves from one shape to another.

EXERCISE 123 *Action Concept*

To get a feeling for moving dynamically as a group, choose four words that suggest four different kinds of actions, such as rise, drop, slash, and surge. The group may choose, for example, to start kneeling in a tight circle, rise slowly as a group to their feet, and suddenly drop one by one to the kneeling position; then, as if chopping a tree down, they may in unison swing their arms and body repeatedly in the same direction, and then as a tight group move forward as if they were a huge wave moving into the beach. Group action, as usual, should arise spontaneously, with no preplanning of details.

EXERCISE 124 *Mood Concept*

To get the feeling for establishing a group mood, select four mood words, such as pensive, fearful, angry, and submissive. As an example, the group, pressed tightly together, might as a group slowly breath in and out as if heaving great sighs, then, still pressed tightly together, move backward as if retreating from some horrible spectre, then with collective courage stamp its feet and move menacingly forward toward the object, only to cower before it and slowly sink to its knees.

4-14

Exercise 125: Expression of a Phrase

EXERCISE 125 *Phrases as a Theme*

By selecting one phrase the group can try to establish a particular idea. The group must try to express the imagery and the concept of the selected phrase *without* discussing what actions are to be done. Remember that you are trying to communicate nonverbally. Keep developing and repeating the action until the group has established the idea *as a group* and not as a number of separate individuals. For example, if the phrase selected is "growing tree," the group might move together and from the floor suggest the growth of the trunk of the tree upward, ending with the hands and arms slowly extending away from the trunk of the tree to suggest branches (figure 4-14). If the phrase is "atomic blast," the group might perhaps crouch very tightly together on the floor, and then slowly rise and spread out and upward in the manner of the rising mushroom of an atomic blast. If the group has difficulty working together, take a

98

few minutes to discuss and plan how these various concepts in movement can be suggested and tried without words.

Combining the idea movement with appropriate sounds can be more challenging and fun. For example, if the phrase is "rainy weather," the members of the working group can individually express *through movement only* the different kinds of sounds heard when it is raining, while the group watching can *make the appropriate sounds*. The sounds can represent anything from a single drip to a pounding storm.

EXERCISE 126 *Situation*

Improvisation with smaller groups can also be based on more complicated ideas or situations. Below are listed a number of suggestions for further bodily exploration. With each improvisation try to react to the situation as you would in real life or as the object would. Try to make each action clear and distinct within the framework of a beginning, a middle, and an end for each gesture and for the scene as a whole. To make the scene more challenging and spontaneous, each person should know only his role in the improvisational situation, so that he is forced to react to whatever the other person is doing without preplanning. The participants should develop an appropriate ending for each scene beyond the basic situation.

a. A piece of gum is stuck to the bottom of your shoe. When you try to get it off, it gets stuck on your fingers, your knee, and elsewhere. A friend, seeing your plight, trys to help, but the two of you eventually get stuck together.

b. Two people enter a dark room from opposite sides of the room. One person is a burglar and the other the owner of the house. Both are trying to find the light switch in the dark, and they both do it at the same time.

c. A person walking late at night on a dark street in a part of town where there have been numerous robberies is suddenly confronted by someone who has been standing in the shadow of a doorway.

d. One person accidentally loses his wallet. The other person finds it and sees that it is full of bills. Meanwhile the first person returns, frantically searching for it.

e. A person enters a restaurant, orders a meal from the waiter, is served, eats, and when presented with the check discovers that he does not have the money with which to pay the check.

f. A young man on the way to his girl friend's house to propose marriage gathers a bouquet of flowers en route which he will present to his girl

99

friend. The girl friend in the meantime prepares for the visit. He arrives at the door, is invited into the house, and presents her with the bouquet of flowers, to which she is allergic.

g. One person impersonates an electric coffee maker and the other person plays someone who really needs a cup of coffee. The coffee lover makes a pot of coffee, pours it, and drinks it.

h. One person plays the conductor of an orchestra, and one or several other persons play all of the musicians in the orchestra, so that when the conductor directs, they play the violin, the kettle drums, the cymbals, and the other instruments. Alternately, one person can be an instrument of some kind, such as a bass fiddle, and the other person can be the instrumentalist. Perhaps sounds as well as body actions will enhance the scene.

i. One person portrays a leaf that is lifted and supported by the action of the wind. The other person plays the wind. Suggest the action without actually lifting the person who plays the leaf.

j. One person plays several pieces of fabric. The other person plays a tailor who threads a needle and sews the fabric together.

BASIC DANCE PATTERNS FOR ONE, TWO, OR A GROUP

Ideally the performing artist should try to study some movement discipline such as fencing, juggling, gymnastics, tai chi ch'uan, or dancing. Each of these disciplines can add immeasurably to body control. Dancing is particularly useful in that it contributes a number of things to the performer's craft that directly relate to his ability to use his body expressively: for example, coordination, quick reflexes, strength, flexibility, movement expressivity, and a sense of line and form.

From time to time a script or an audition may require that you dance. Although there is no substitute for the actual dance training, moving in some kind of formal dance pattern will give you a feel for dancing. Although it is outside the scope of this book to present specific dances, a few exercises are presented below that may be helpful in developing your coordination and a feel for some of the basic patterns used in period dances.

You can do the following exercises with a group, alone, or with a partner. Working with a partner or a group is valuable in that it will force you to coordinate with others. In working with a partner, perform the same movement patterns while standing side by side, facing in the same direction, and lightly holding one hand of the partner with both arms extended to the side at shoulder height. Then do the same movement patterns facing one another with the boy's right hand on the girl's waist, the girl's left hand on the boy's shoulder, and the free hands joined and held to the side. The girl's movements are the opposite of the boy's; for example, if the boy steps forward on the right foot, the girl steps back on the left foot.

In all of the following exercises, step or move onto the right foot first (if you are the boy), concluding the sequence by bringing the feet together. Repeat the exercise starting on the opposite foot and reversing the direction of any turns. Establish where the audience is so that each time you have a reference point to which to return. Start each sequence facing forward. To develop the ability to move to music you may want to select a piece of music with a strong beat and do the steps to it.

EXERCISE 127 *Walk-Run*

Take four steps forward, each step on one count. Next, without a stop, make a one-quarter turn to the left (90°) and take eight steps forward, allowing yourself only four counts. In other words, double the steps for each count. Without stopping, continue the floor pattern by making a three-quarter circle turn to your left (270°) that brings you back to the beginning position. As you make the three-quarter turn, bend the knees slightly and increase the speed of the walk until you break into a run. Make the three-quarter turn in eight counts and use as many steps as you wish. Bring the feet together and continue immediately onto the left foot, repeating the turns to the right. Try to keep the movement smooth and rhythmical, with a minimum of vertical rise and fall in the body. The same sequence can be repeated stepping backwards, ideally without looking backwards, so that you develop a sense of where you are in the room by using your peripheral vision (that is, what you can see at each side of your body without shifting your eyes from side to side).

EXERCISE 128 *Open-Closed Steps in a Square*

Allowing one count for each step, step forward on the right foot and bring the left foot into place next to the right, ending with the weight on both feet. Next, step to the right side on the right foot and bring the left foot next to the right, ending with the weight on both feet. Next, step back on the right foot and bring the left foot to the right foot, ending with the weight on both feet. Conclude the sequence, stepping to the left on the left foot and bringing the right foot next to the left, ending with the weight on both feet. At the conclusion of the sequence your steps should have traced a square on the floor. Repeat the sequence four times and then repeat it four more times to the other side, starting on the left foot.

101

EXERCISE 129 *Grapevine Step*

Allowing yourself one count for each step, step to the side on the right foot, cross the free foot in front of the supporting foot and step onto it, step sideward again with the beginning foot, and cross the free foot in back of the supporting foot and step onto it. Continue the steps and crosses for seven counts, and then on the eighth count bring the feet together with the weight on both feet. Repeat to the other side. Repeat again to the original side in only four counts but with eight steps, so that the steps are done double-time to the music. Repeat the four-count sequence to the opposite side. Try to keep the steps smooth and rhythmical. It may be helpful to bend the legs as you step across one of them.

EXERCISE 130 *Step-Ball-Change*

In a three-count sequence, step forward on the right foot on the count of one, on the second count bring the left foot up to the right foot and step onto the ball of the left foot, and then on the count of three step forward onto the right foot. Repeat the sequence to the other side, stepping onto the left foot, onto the ball of the right foot, and onto the left foot. Repeat, varying the direction stepped and the facing direction.

EXERCISE 131 *Slide-Cut-Step*

In a three-count sequence, slide the right foot forward along the floor, ending with the weight on it on count one. Quickly bring the free left foot forward as if to kick the supporting foot out of the way, take the weight on the left foot as you lift the supporting right foot from the floor on count two, and step in place on the free right foot on count three. Repeat the action forward on the left foot, the right foot closing into the left and taking weight, and concluding by stepping onto the left. Repeat the action to each side and to the back.

EXERCISE 132 *Pivotal Changes of Direction*

Starting on the right foot, in four counts take four steps forward. On the fourth step, pivot on the supporting left foot one-quarter turn to your left (90°). Continue the sequence to each side of the room until you are facing forward again. Next, without stopping continue the four-count step pattern forward, taking a one-*half* pivot turn to your left (180°) on the fourth count, so that you face in the opposite direction from which you started. Continue the walk pattern, taking a one-half pivot turn to your left on the fourth step, so that you face front again. Practice the exercise turning to the right, as well. At first, practice the quarter-turn and the half-turn separately. Then try to combine them without stopping. When you are facing a partner, there will have to be some adjustment on the last step so that you face opposite one another.

EXERCISE 133 *Full Pivot Turn*

Starting on the right foot, take four steps in place, making a full pivot turn to the left (360°) pulling the left shoulder back on the fourth step. Repeat the pattern, making a full pivot turn to the right on the fourth step. Stop and repeat the sequence, starting on the left foot and turning first to the right and then to the left. If working with a partner, drop the arms and turn independently of one another.

EXERCISE 134 *Vertical Bending and Rising Actions*

a. Down, Up, Up (Waltz) Allow three counts for the sequence. Step onto the right foot with the knee bent; next, step onto the ball of the left foot with the knee straight; and next, step onto the ball of the right foot with the knee straight. Repeat the sequence on the left. Continue the action so that on the count of one of every new three-count sequence the free leg is bending as you step on to it. Try the sequence changing directions and traveling.

b. Down, Up, Up, Down Allow four counts for the sequence. Step forward onto the right foot, bending the knee. Step forward onto the ball of the left foot with the knee straight. Place the right foot beside the other foot. With the weight on both feet, lower the heels to the floor, with the knees remaining straight. Repeat the sequence, alternating feet and facing directions.

c. Down, Up, Up, Down (in Place) Do this exercise slowly, taking one count for each new action. The toes are facing straight ahead, the weight is evenly distributed between the two feet, and the body is pulled vertically upward in a good posture. Bend both knees, keeping the heels on the floor. Next, straighten the knees. Rise up onto the balls of the feet, and then end by lowering the heels to the floor, keeping the knees straight. Repeat the sequence several times.

EXERCISE 135 *Step, Tap; Step, Point Foot*

Start with the toes pointed forward and the weight evenly distributed on both feet.

a. Taking one count for each new movement, step forward onto the

103

right foot, and then bring the left foot next to it, tapping the ball of the left foot on the floor. Repeat the action, stepping onto the left foot and tapping the right foot.

b. Repeat the action again, starting with the right foot, but this time with the legs rotated outward from the hip sockets, tapping the ball of the free foot slightly forward at the instep of the supporting foot. Repeat with the left foot starting.

c. Repeat the sequence from the top, but instead of bringing the feet together for the tapping action, extend the tapping foot (left) forward of the right leg approximately one foot. Again tap the ball of the left foot on the floor. Repeat with the left foot starting.

d. Repeat the sequence from the top but instead of tapping the ball of the foot on the floor, tap the tip of the toe with the leg extended forward, the knee straight, and the instep of the foot stretched downward ("pointed").

EXERCISE 136 *Step, Brush (or Swing) of the Leg*

a. Allowing one count for each action, take three steps forward, starting on the right foot. On the fourth count lightly brush the ball of the left foot on the floor as the left leg swings forward. End with the left leg straight and the toes pointed downward. At the same time that the left leg swings forward, bend the supporting leg. Repeat the action starting with the left foot, so that the right leg brushes forward.

b. Try the sequence with no inward or outward rotation of the legs and also with outward rotation of the legs from the hip sockets.

c. Try the same sequence brushing the free leg to either side or to the back. In each exercise, after stepping forward bring the free foot to the instep of the supporting foot in order to slide the ball of the foot on the floor as the leg is being extended away from the body.

d. For variety, try the same action stepping only once and brushing the free foot in some direction.

e. Try the sequence again, keeping the supporting knee straight while brushing.

EXERCISE 137 *Steps Rising off the Ground*

a. Hop on One Foot In a three-count sequence, step on the right foot and hop from the right foot into the air, landing on the same foot (the step, hop, and landing all done in one count). Next step on the left foot for one count without a hop and then on the right foot for one count

104

without a hop. Repeat the sequence hopping off the left foot. The sequence is made up of step/hop, step, step; step/hop, step, step. It can be counted "and one, two, three; and one, two, three." The first step and hop have to be done quickly in *one* count. After you have the feel of the hop action, try moving in various directions and turning.

b. Skip (Alternating Hops)   Step on the right foot, hop off that foot into the air, and land on the same foot. Repeat the hop on the left foot and keep repeating the hop off alternate feet. The skip is always done in an uneven rhythm with a sense of lightness to the movement. Each step-hop action can be done in one count, and can be counted "and one, and two," and so forth. Vary the skips by traveling in different directions and while turning.

c. Jump from Two Feet to Two Feet   In a three-count sequence, step forward on the right foot on count one. On count two bring the left foot beside the right foot and bend the knees, keeping both feet flat on the floor and the body pulled up in a good posture. Next, jump quickly into the air from both feet, straightening the knees and trying to point the tips of the toes toward the floor. On the third count land with the knees bent. Try to land on the ball of the foot and quickly lower the heel to the floor (a rolling action through the foot). Repeat the three-count sequence, stepping forward on the left foot. Vary the pattern by stepping into various directions while continuing to face forward. Vary the pattern further by stepping into various directions at the same time that you change the facing directions. Vary it still further by traveling in a new direction while jumping into the air.

d. Leap from One Foot to the Other Foot   A leap is a jump into the air off of one foot and a landing on the opposite foot. Use a three-count sequence, with one count for each action. On the first count, step forward on the right foot, bending the knee. On the second count, swing the left foot from the back to the front, at the same time jumping off the supporting foot and landing on the left foot on count three. Continue the action, stepping onto the right foot, swinging, and leaping onto the left foot. The exercise will be easier if done quickly. Vary the sequence by stepping or leaping into various directions, or by taking more than one step before the leap action and by starting the sequence stepping onto the left and leaping onto the right foot.

e. Jump from One Foot to Two Feet   In a two-count sequence, step forward onto the right foot, bending the knee on the first count. On the second count, brush the left leg from the back to the front while jumping into the air from the right foot, and then land on both feet (rather than on one foot as above). Repeat the action, stepping onto the left foot, jumping

105

off the left foot and landing on both feet. This is sometimes referred to as an assemblé. Vary the sequence by stepping or turning into various directions.

f. Jump from Two Feet to One Foot In a four-count sequence, start with the weight on both feet. Bend the knees and jump into the air on the first count, landing on the right foot on the second count. On the third count, put the left foot down and jump into the air from both feet. On the fourth count, land on the left foot. This is sometimes referred to as a sissonne. Vary the pattern by traveling or facing in a new direction when jumping.

In the above exercises no attempt has been made to tie the movement sequences to a specific dance rhythm or dance form, but rather to present some of the common kinds of movements that you might encounter in a period dance. The value in doing the exercises is to help develop your coordination and rhythm as well as to establish in your body the kinesthetic sense for dance movement and in your mind a way of distinguishing various movement patterns.

The above exercises can be done by yourself or with a partner. You may want to try them with a group of from six to eight people and do several of the patterns together as a group dance. The value in doing them as a group is to develop your capacity to move and act as a group member rather than as an individual. Working with a group requires concentration on keeping a regular beat, judging yourself and the space around you in relation to other people, and reacting quickly and rhythmically to a change of hands, arms, space, or movement patterns. The shared experience can also be fun if you enter into it with a joyful spirit.

If you decide to work with a group, choose a group leader who will count or clap out the rhythm for you. Better still, find some spirited, rhythmical music to dance to, such as square dance music. Decide ahead what the sequence of steps will be, what foot you will start on, and how many repetitions of each step you will perform.

One of the easiest group patterns is the circle, in which you face into the circle and join hands with the person on either side of you. Another common pattern is two lines of people facing each other. The circle can separate into the two lines or the two lines can join together into the circle.

An example of a group effort is presented here as the basis for further exploration.

EXERCISE 138

Six or eight people hold hands in a circle. Stepping to the side on the

106

right foot, do the "grapevine," moving to the right for eight counts and ending with the weight on both feet on the eighth count. Repeat the action, stepping on the left foot and moving to the left for another eight counts. Next, starting on the right foot, do the "skip step," facing and traveling to the right for four counts, facing and traveling to the left for four counts, moving forward toward the center of the circle and raising the arms for four counts, and then lowering the arms and skipping backward for four counts, separating into two straight lines facing one another. Next do plain walks, starting on the right foot: raising the arms, walk four steps toward the other line; then, lowering the arms, walk four steps backward away from the other line; and then, raising the arms, walk four more steps toward the other line. Dropping the hands of your partners, pass through the line of the other group (members of the two groups passing by each other alternately). Next, in the four remaining counts of the walk action, turn to the right and walk back into a circle, joining hands. Using this example as a basis, explore other group stage patterns and steps.

EXERCISE 139

Other exercises that are beneficial in developing body awareness and control for dance or movement in general are based on rhythm. Decide on a set number of counts (three, four, six, or twelve, for example) that will be repeated a number of times as musical or movement phrases. Establish the rhythm by clapping the sequence. Clap in the same speed throughout and heavily accent the first count in each new set. If, for example, you have decided on sets made up of four counts to be repeated a total of four times, clap the first set, step in place for the second set, move the whole body freely through space on the third set, and move a different part of the body on the fourth set. Then repeat from the beginning.

The purpose of this exercise is to move spontaneously as well as to develop a sense of rhythm. The tendency for many people is to speed up the rhythm rather than keeping it at a constant pace. You should be able as a group to establish an audible steady rhythm with the claps and steps, keep the rhythm steady when moving silently, and start the claps again at exactly the same time.

Each sequence should be a distinct "movement phrase." On the first count make the clap or movement very clear, so that someone watching or listening will know that a new sequence is beginning. Once you feel comfortable moving rhythmically with each count, explore more complex rhythms of clapping and moving. For example, using a four-count

107

sequence, every so often double the speed of the count so that there are two claps in the amount of time that you are ordinarily clapping one beat. The sequence might be 1, 2 (3 and), 4. Here the third count is doubled so that there are two quick claps instead of one. One sequence of alternating four-bar rhythms is: 1, 2, (3 and), 4; (1 and), 2, 3, 4; 1, 2, 3 (4 and); and (1 and), 2, (3 and), 4, which is then repeated from the top. Each four-count set should take the same amount of time whether a count is doubled or not.

EXERCISE 140

Another exercise in rhythm and coordination is to walk in one rhythm while moving the arms in another rhythm. For example, count 1, 2, 3, 4; (1 and), (2 and), (3 and), (4 and) while walking in place. Take one step for each single and double count (a steady four-count). Meanwhile, lift the arms smoothly forward of the shoulder in four counts, over the head in another four counts, to the back at shoulder level in four counts, and down beside the body in another four counts.

Explore other varieties of this exercise by moving the arms smoothly and slowly in other patterns while stepping in more complex rhythms.

Explore the problems of keeping the rhythm with some other part of the body (by nodding the head, for example, or pushing the hips to the side) while moving the leg, arm, or another part of the body in a sustained motion or complex rhythm.

EXERCISE 141

a. Using the mirror image idea, the group leader does a simple, rhythmical movement phrase which everyone else repeats.

b. The group is arranged in a circle. The first person sets up a four-count rhythm by clapping the hands, performing movement of some kind, or making sounds with the voice. The person next to him repeats the rhythm verbatim, and it is repeated in turn around the circle.

c. Next, the first person establishes the first four-count movement or sound phrase, and the next person repeats it and adds to it his own four-count action or sound. Each person in turn repeats what has been established and adds his own phrase, until the pattern reaches the original "sender," who ideally can repeat the entire pattern. At first you may want to limit expression to one device, such as clapping or vocal sounds. Later part of the challenge and fun is to repeat four-count sequences that include clapping, movements, and vocal sounds. The purpose of the exercise is to help you to concentrate on seeing, hearing, and repeating what you have seen and heard.

There will be occasions when you are required to participate in a stage fight. Certain kinds of fights require handling and using weapons such as rapiers or broadswords in a very specific and skilled manner. These are specialized skills that require training under the watchful eye of an instructor or director. Often university and college physical education departments offer courses in fencing or judo. Along with dancing, these studies can add enormously to your ability to move with skill, grace, balance, rhythm, and control.

Although it is outside of the scope of this book to deal with any particular fighting technique, a few basic principles may be helpful to you.

Any fight consists basically of defensive and offensive moves. In defense one tries to protect the most vital areas of the body and conversely in offense one tries to break through the defense of the other person to render a blow.

In defense the most important areas of the body to be protected are the head, the neck, the abdomen, and the left and right sides of the body from the hip to the rib cage. The defensive fighting posture must be such that these areas are protected and the smallest possible area is vulnerable to the attacker. To block an attack to any of these areas requires a specific kind of movement. For example, in boxing, if a person is right-handed and his opponent strikes toward the head, the right arm is lifted upward to protect the head. If the strike is toward the right side of the neck, the right arm deflects the blow to the right; and if the strike is toward the left side of the neck, the person blocks the opponent's strike by moving his right arm across his body sideways to the left. An attack to the right side of the trunk is repulsed by moving the right arm downward and to the right; and the left side of the body is protected by moving the right arm to the left side across the body and toward the back. The abdomen is protected by shielding the abdomen and counterattacking the aggressor. The principles of defensive blocking are the same when weapons are employed.

An organized group usually has a similar aim of maintaining the line of defense or conversely of breaking through the defenses of the opponent group. When the line of defense is broken, a general melee of hand-to-hand combat may ensue.

Throughout history, of course, many different kinds of weapons have been used in hand-to-hand combat. The kind of weapon used will affect the strategy of defense and offense, the manner of moving, and the basic stance employed. The shape, size, weight, and destructive design of the weapon will determine how it is held and manipulated. Weapons may be designed to pierce, slash, or crush the opponent. For example, a dagger might be used to pierce, a sword to slash, or a ball and chain to crush the opponent.

There are two basic kinds of dynamic action that take place in a fight: punching or thrusting actions and slashing actions. The punch follows a straight line through space while the slashing action generally follows a circular path in space. Either action can be used to block the opponent's action or to deliver a blow. Either action can move in any direction in space and with varying degrees of energy, depending on its purpose. For example, in sparring with a partner the boxer might use a series of rapid jabs, delivered with a minimum of energy in a series of straight lines through space. Their purpose would be to break down the defense of the opponent. When an opening occurs the fighter could deliver a series of punches with a maximum amount of energy and perhaps follow them up by a slashing action that delivers a blow to the chin from below or from the side.

Any kind of gestural action goes through a specific series of phases: the body's preparation for the action, the performance of the action, the followthrough, and the return of the body to a stance for preparation to either attack or defend. The person acted against goes through a similar series of phases: body preparation to ward off the blow, defensive action, followthrough, and recovery. If the aggressor scores a hit, the followthrough for both participants will be different than if the blow is warded off.

The kind of energy used and the spatial directions moved through in a fight are important because the stage fight must look spontaneous to the audience and yet be controlled so that neither opponent is accidentally injured. A fight must be carefully and slowly rehearsed so that both participants know exactly which moves to execute, how much energy to put behind an action, and how to stop an action just short of contact. To make it appear that contact has taken place, the fight must be done with split-second timing between the participants. The blow appears to be delivered if the receiver reacts appropriately to the apparent force and direction of the blow. For example, if someone were hit in the stomach his torso would collapse forward, while if he were hit under the chin his head would snap back and his body would be thrown backward.

Fights on stage serve a variety of dramatic purposes. A fight may serve to further the plot, reveal something about a character, build dramatic intensity, or simply provide comic relief. The purpose of the fight will influence how the fight is performed. A man and a woman will fight differently, as will trained and untrained combatants. The styles of a

110

fight may reveal something about a character: for example, that he is a coward, a bully, or courageous. The fight may be done realistically with great intensity or exaggerated for comic effect. A large fight may be a brawl, such as a typical western saloon fight, or precision operation, like the fighting of a Roman army.

Whatever the purpose of the fight, it still follows a basic pattern of action. The cycle of a fight is somewhat like the cycle of unbridled anger. The anger starts slowly, builds until it explodes in activity, and then expends itself and is slowly brought back under control. The following exercises are presented to help you experience physically some of the basic principles of attack and defense. Aside from their practical value, they are another method of learning to relate to a partner and to move sensitively in response to another person. They also can help establish a sense of timing and dynamics.

4-15

Exercise 142:
Stick Fighting

EXERCISE 142 (*Figure 4-15*)

Find an object such as a yardstick or, preferably, something tubular of about the same length. Place one hand on each end of the stick and hold it in front of you parallel to the line of your shoulders. Face your partner (who also has such a stick) at approximately arm's length and take turns using the stick defensively or offensively with your partner. Make all movements in slow motion, expending a minimum of energy so as to give you both sufficient time to insure that no one is hurt.

a. The first time you do the exercise concentrate on offensive actions directed toward the head, the sides of the neck, the sides of the trunk,

111

and the abdomen. In each case the actions should be blocked by the person on the defense.

b. Repeat the actions in slow motion, imagining that the offensive action has not been blocked and that your partner receives the blow. In slow motion try to use the energy that you would need to strike someone and the kind of followthrough your gesture would require. The person receiving the imaginary blow in slow motion should respond appropriately to the kind of blow received, concentrating on the followthrough and recovery. When you are working in slow motion, make certain that a logical followthrough takes place. You must imagine what would happen to your balance, how speed would force your actions in a certain path, and other such matters.

c. Repeat the exercise again in slow motion, emphasizing the idea that both partners are equally matched, so that you take turns advancing on the offensive and retreating on the defensive.

EXERCISE 143

a. Discard the sticks and perform an imaginary fistfight in slow motion. Concentrate on the timing that is necessary to make each blow given or received appear real. Be sure that you do not make contact with the other person, but come close enough that you appear to have contacted him. Try the same actions at a somewhat faster pace when you feel you have the ability to act or react with the other person safely.

b. Try the same kind of fight with imaginary swords or some other imaginary weapon, such as a dagger, club, or two-handed sword.

4-16

Exercise 145:
Faint Fall to the Side

Devise a scene of some kind and try to establish some kind of characterization with the fight. For example, a bully comes into the room in a menacing manner and threatens a number of people. One person accepts the challenge and, after a few punches, the bully reveals himself to be a coward and backs off.

STAGE FALLS

Whether or not you are ever required to fall on stage, practicing a few basic stage falls will give you some ability to move with control to the floor. Basically a complete fall is made up of a loss of balance, an effort to recover the balance, and a fall to the floor. In a performance situation, complete control over your body is necessary to prevent injury. The context in which the fall takes place will affect the dynamic energy requirement and the speed of the fall. For example, a fainting fall would be slower than a fall resulting from being thrown to the floor or a fall that results from tripping over something. In a stage fall there is always the element of surprise for the audience; therefore they do not have time to analyze your technique, provided that the fall flows without interruption from one position to the next.

The three following falls (to side, the back, and the front) can be modified to fit the dramatic situation by changing the speed and the dynamics.

EXERCISE 145 *Faint Fall* (*Figure 4-16*)

The simplest stage fall is the "faint" fall. As the name "faint fall" suggests, there may be at first a sense of light-headedness and a feeling of floating projected, followed by a complete relaxation of the seemingly unconscious body as it falls toward the floor. The technique of the fall is relatively easy and quite safe if practiced carefully. If you are going to fall to the right side, take the weight on the left foot and move the right foot back. Keeping the torso erect, lower yourself in a deep knee bend

towards the floor until you can place the right knee on the floor in a kneeling position. Never allow yourself to drop onto the knee. Shift your weight backward until you are almost sitting on the bent right leg. Shift your weight to the right so that the fleshy part of your upper thigh and buttocks touches the floor. With practice you may not have to use your arms for additional support, but in the beginning place both hands on the floor for support and continue to lower yourself toward the floor, taking weight on the right side of the body. Each part of the action should flow smoothly and a completely relaxed quality must be projected. At first practice kneeling until that can be performed with ease and then practice falling to the side from the kneeling position. Then combine the two actions and try to practice the fall as if you were actually fainting.

A faint fall to the back is done in a similar manner. Take the weight on the left foot, move the right foot back, and lower yourself to a kneeling position with the right knee on the floor. In order to fall backward, turn the right leg outward in the hip socket. Shift your weight toward the back until your buttocks come in contact with the floor and then roll yourself vertabra by vertabra toward the floor until the whole back is on the floor.

EXERCISE 146 *Stumble Fall*

In the stumble fall, the balance may be temporarily lost and then re-covered before you hit the floor. On occassion, however, it may be effective or necessary to fall all the way to the floor. After establishing the idea that you are off balance, step forward on one foot, bend the torso forward to a right angle, bend the legs, and place the hands on the floor in front of you. Lower yourself to the floor with your arms. Slide your right arm along the floor over your head and roll over onto your back.

EXERCISE 147 *Combining Movements*

Practice the three falls with different kinds of motivation: for example, in response to fainting, stumbling, and being thrown to the floor in an imaginary struggle. Be aware of the kind of impulse or energy needed to start the fall and the energy the fall itself requires.

After you have practiced by yourself, work with a partner who in a struggle or fight supposedly causes you to lose your balance and fall. For example, perhaps you are hit with an uppercut under the chin. Your head and body are thrown backwards. Perhaps to try regaining your balance you make a complete turn of your body, which leads to a back-fall. In an exaggerated version the backfall may lead to a backward somersault. You may end by lying on your back or landing on your feet and preparing for a counterattack.

It is probably best to practice these falls in slow motion at first. Later you may care to speed up the action. The timing and the dynamics must be so well worked out that a real loss of balance never takes place and yet the illusion is complete.

Men are like rivers: the water is the same in each,
and alike in all; but every river is narrow here,
is more rapid there, here slower,
there broader, now clear, now cold, now dull, now warm.

LEO TOLSTOY

5 Looking at the Movement of Individuals

In life every human being is unique. Individual movement patterns are contingent upon many factors, including age, sex, body build, body condition, personality, and environment. As a person lives out his life, his attitudes and movement habits are reshaped in part by the changes in his body and in part by the events that surround him.

The attitudes of the family, the society, and the state all influence the way people behave at various ages. Education, financial resources, occupational skills, social class, peer groups, society's mores, and climate all play their part in determining the way people perceive the world and themselves and hence also affect their way of moving.

The following discussion provides approaches for observing the movement of human beings. If they are to be useful to you, you must

116

confirm or refute them by constant observation of yourself and others in all kinds of situations. You must also try regularly to move in the ways that you observe under these varying circumstances, at various times of the day, the month, and the year.

HABITUAL MOVEMENT PATTERNS

As a person goes through life he develops a habitual and characteristic way of moving. His movements become a personal body language that can be understood and analyzed in the same way as his spoken language. Although no two people move in exactly the same way, there are similarities as well as differences, which you can train yourself to observe. With careful study you will begin to understand that movement cannot be rigidly categorized by age, class, or sex. Once you have trained yourself to really look at the way people move, you can avoid movement clichés when developing a role. For example, the fact that a person has lived for seventy years does not necessarily mean that he will have stiff joints and walk with shuffling steps. You will also find that comparing the ways in which individuals move is an exciting adventure in itself. The time spent in observation of others will be worth the effort.

Learning to analyze and understand the meaning of body movements has, particularly in recent years, become a serious scientific endeavor. The work of people like Laban, Birdwhistell, Wolff, and others is systematizing for study what performing artists have always known or at least practiced intuitively.

Many scientists are of the opinion that there are no meaningless movements. The meanings of the individual movements, however, can be known only within the framework of a movement pattern and dramatic context. Looking at movement the way a scientist does can be valuable for artistic recreation. Scientists often break a person's total movement pattern down into isolated elements noting the frequency and dynamic characteristics of certain kinds of movements. In developing a characterization, you must eventually decide upon specific kinds of action and join them together into a pattern expressive of your view of the character. Like the scientist you can analyze movement to reveal a communicative pattern, and then select isolated elements of movement to create character in a specific dramatic context.

EXERCISE 148

From the categories below (carriage; use of the body, space, time, harmony, and dynamics; and body image), select one in which you will care-

117

fully observe how some individual moves. Later do the same with the other categories. In the beginning it is easier to limit yourself to one area of concentration so that you get a feeling of what to look for in movement. You should also become aware of how you move in these areas as well.

Body Carriage. When walking, standing, or sitting, is the body stretched vertically upward, pressed downward, widened to the side as if expanding, laterally compressed on itself as if shrinking, or carried somewhat forward, backward, or sideward? Does there seem to be tension or stiffness in any part of the body, particularly in the shoulders, neck, or torso?

Use of the Body. Does the person tend to use one side of the body more than the other? For example, does he frequently gesture with the right arm only? Does the person move one area of the body much more than the rest? For example, some people keep the torso and head rigidly in place and make most of their gestures with the lower arms or hands. Does the individual frequently repeat a certain gesture? For example, does he twist his mouth in a certain way when talking, nod his head, or move or hold his hands, feet, arms, in a characteristic way?

Use of Space. Are shifts of the body weight or gestures of the head, torso, arms, or legs frequently made in the same direction? Are the gestures directed more often toward or away from the person making them? Are the gestures fully extended outward in space or held back? Do the gestures appear complete or incomplete? Do the gestures move in straight or curved paths through space? Does the person tend to make characteristic patterns of gesture in space?

Use of Time. Are the movements primarily fast or slow? Do the movements have a repeated rhythm or does the rhythm change frequently? Do the movements suggest that the person is fighting against time or does he seem to have ample time? Does the person combine many different movement patterns rapidly or does he use only a few over a long period of time?

Use of Harmony or Disharmony. Do the movements appear to flow smoothly or are they frequently disjointed or interrupted? Do the movements of the various parts of the body relate to one another harmoniously or does the movement of some part of the body seem to fight against the basic movement pattern? Are the movements made easily and gracefully or do they seem tight and restricted? Do the movements seem heavy and slow, as if performed against the force of gravity, or do they seem light and quick?

Use of Dynamics. Does the person repeatedly make sharp or percussive gestures, such as punching the air with the fist, pointing and thrusting a finger to make a point, drumming the fingers, or tapping the feet? Do the gestures or the body seem to be pushing or pulling through space? Does the person make gestures that slash through the air? Are his movements generally sustained or is there frequent hesitation before, during, or after a gesture? Do his gestures make repetitive circular paths in space, as if he were stirring up the air? Do they cause the body or any of its parts, such as the head or legs, to swing or sway?

Adaptation to Others. Does the person lean forward to others, as if relating to them, does he hold himself back, as if retreating? Do his movements seem responsive or unresponsive to others? Does the individual tend to take on the gestures and movements of those present? Are the person's movement rhythms and mood the same as those of the group or different? Are the movements appropriate or inappropriate to the situation and to what is being said? Do the movements suggest that the person is comfortable or uncomfortable with the group? Are the movements disturbed and agitated or are they composed and relaxed? Do the movements of the person suggest that he is involved in the group or distracted by activities or people outside of the group?

BODY IMAGE

A person cannot escape the kind of body with which he is born. However, exercise, diet, make-up, selection of clothing, body carriage, and attitude can alter the body's appearance, especially on the stage.

Everyone has an image of what he looks like. People see themselves, for example, as tall, short, thin, fat, young, old, shapely, handsome, graceful, unattractive, or clumsy. This image influences their thinking, their personality, and hence their way of moving. Sometimes the image is quite different from what a person really looks like; nevertheless, it is often true that what is strongly believed in and visualized long enough, whether accurate or not, eventually produces results. So it is with a person's body image, which governs his movement pattern as much as his actual physical structure.

When observing someone, you should note not only the kind of body the individual has but also the image he appears to have of his body and the ways in which his movements suggest his self-image. Psychiatrists relate that acceptance of one's body is vital in developing a secure personality. Children as well as adults may have severe personality disturbances and an inability to adapt socially if they cannot accept their body image.

Later when preparing for a role you can determine how to alter

119

your own postures, stances, and movements in order to project the kind of body image that is right physically and psychologically for the character being portrayed.

PERSONALITY AND BODY TYPES

For centuries people have made attempts to categorize each other in order to study and understand their behavior. Fortunately, human beings are too complex to be rated arbitrarily and finally on some selected scale. Nevertheless, as long as it is understood that individuals are not stereotypes, categorizing people by personality or body type, for example, can be useful in gaining general insights.

Foreshadowing the modern science of endocrinology was the early attempt of Hippocrates, around 400 B.C., to categorize individuals by the type of fluid thought prominent in their bodies. The four fluids considered were the blood, black bile, yellow bile, and phlegm. Each was thought to produce a characteristic temperament: sanguine, melancholic, choleric, or phlegmatic. Each was thought to produce a specific mood and general level of activity. These types are sometimes referred to in dramatic literature. More recently another theory categorized men by their physique.

Three basic body types were distinguished: the ectomorphic, mesomorphic, and endomorphic (figure 5-1).

Ectomorphic. The body is characterized by long limbs; a narrow

5-1

Body Types

Ectomorphic   Mesomorphic   Endomorphic

120

trunk, shoulders and hips; a long, narrow face; stooping shoulders; delicate skin; fine hair; and a sensitive nervous system. In personality and temperament the ectomorph is characterized as hypersensitive, fearing groups, needing solitude, serious, dull, and phlegmatic.

Mesomorphic. The build is characterized as athletic, with well-proportioned limbs, good muscular development, broad shoulders and trunk, and narrow hips. Mesomorphs are energetic, like exercise, and are direct in manner. In temperament they tend to be moderate, midway between ectomorph and endomorph.

Endomorphic. The body build is characterized by short limbs, broad trunk, a round face, and a tendency to gain weight in later years. The temperament varies between extremes of gaiety and depression. The endomorph loves to eat, seeks body comforts, and is sociable.

C. G. Jung introduced the psychological terms "introversion" and "extroversion" as aids in tracing basic personality trends. They are merely extremes on a scale and do not indicate two actual types. Introversion is characterized by a tendency toward the inward projection of interest and attention, daydreaming, introspection, contemplation, and deliberation in making decisions. In times of emotional stress the introvert tends to withdraw into himself. Extroversion is characterized by sociability, an outgoing interest and attention, and a tendency toward impulsive action and decision. In times of emotional stress the extrovert tends to lose himself among people.

Considered by some as a more accurate guide to personality assessment is the study of an individual's specific behavior traits, the kind of personality pattern the traits indicate, and how the individual uses these traits in his environment under varying circumstances. One might consider how the individual expressed himself in social adaptability, emotional control, conformity, intellectual inquiry, and confidence in self-expression. With this system a unique personality structure could be charted that showed a unity between expressive movement and personality. The individual then would be seen to have a consistent pattern of behavior, for example, in the way he walks and talks, and in his handwriting.

Many of our misconceptions that lead to stereotypes are based on the false assumption that we can rigidly categorize people. For example, it is sometimes assumed that all fat people are jolly or all short people are aggressive. These kinds of misconceptions can lead also to stereotype role playing in the performing arts. Recognizing the limitation of categorizing people, you can nevertheless use some of the ideas above as a springboard for more careful observation of people in order to avoid conventional characterization.

EXERCISE 149

Although psychologists ordinarily do not define personality types by body type, it can be a basis for observing people. Find a number of people who fit into one of the three body types and try to predict their behavior on that basis. Do you find any correlation? If not, how *do* they move, and what influence has their basic body build on this movement? For example, does a given tall person move confidently, carrying his body erectly, or does he tend to shrink down in an attempt at being inconspicuous? Does a particular short person appear tall because he acts and carries himself assertively, or does he seem shy and retiring? Does a certain fat person move slowly and heavily or gracefully and lightly? As mentioned above, a person's self-image can affect his movements as much as his basic body type. Thus carriage may reveal much to you about a person's attitudes and general state of mind.

EXERCISE 150

While the terms "extrovert" and "introvert" are useful, they should be used only as a springboard to more careful distinctions among personality types.

a. Observe the movements and behavior of others and classify them as to introversion or extroversion. Is your decision based on movements or personality?

b. Notice those who seem to be the natural leaders and natural followers in any group. Whom does the group seek out, listen to, and eventually follow? What is it about their personalities and their movements that attracts people to them? Do their movements suggest self-assurance?

c. Try to determine whether a person's extroverted or introverted behavior might be a shield for private feelings. For example, is the young man who struts, swaggers, brags a lot, and talks loudly covering up a feeling of inferiority, or is the young, attractive girl who appears "stuck-up" actually revealing that she is shy and modest?

EXERCISE 151

Listen to conversations to find out what kinds of subjects excite different people. Observe how they express their interest: are they intensely involved, angered, delighted? Observe how their body movements reveal their feelings about the subject. Notice the difference in their body movements when they are talking and when they are listening. Notice how they prepare with their movements when they want to say something and how they signal with their movements when they are ready to let another person speak. Notice how their body "listens" when they are interested in what is being said or how they withdraw their body when they are uninterested. Notice the shifts in their body position when shifts in the conservation take place. Try to determine if their movements during conversation reveal anything about their basic personality, self-image, or attitudes toward others who are present.

EXERCISE 152

Observe one or two friends that you know well to see if there is a correlation between their behavior and movement patterns under different social situations in their adaptation to others, their conformity to group pressure, and their manner in self-expression. Try to determine if they have a characteristic way of self-expression and consistent movement patterns in walking, talking, and eating, for example.

AGE

When asked to portray characters of a definite age, students often fall into stereotyped movement patterns. Such patterns exist not only for the old but also for the very young and for students themselves. An old person, for example, is often characterized as shaky-kneed, bent over at the waist, and supported by a two-foot cane. A little reflection on public figures and acquaintances will show the falseness of such general conclusions. Recent research indicates that mans most productive years occupationally are after fifty. A look at the ages of the great statesmen and artists of the world, and how they move, think, and act, will give some idea of the dynamism that is possible with advancing years.

Individuals are as unique in the aging process as in other ways. This process is affected at least as much by attitude, habits of mind, physical activity, and general health as by chronological age. Illness, lack of exercise, a poor diet, changes in attitude and mental habits, and muscular atrophy, for example, can cause premature restrictions in body movement, just as other conditions will delay such restrictions.

In thinking of the age of the character you will play, therefore, you must take into consideration not only his chronological age but also his mental age, his general outlook on life, the circumstances of his life, and his patterns of emotional, intellectual, and physical reaction.

Physiologists currently divide the life span of man into five ages: infancy to seventh year, childhood to fourteenth year, youth to twenty-one, adult to fifty, and old age. Historically these divisions have changed

123

with changes in expected longevity and new viewpoints regarding people's capabilities at various ages. For example, in the Elizabethan period what is currently classified as the period of childhood often saw people married, and old age had a much earlier onset.

Some general observations on movement patterns of the young and old may be helpful to you in establishing age characterizations. As always, remember that these comments are not intended to produce stereotypes.

The Infant. The infant makes random movements of the limbs, is unable to focus his eyes steadily on an object, is unable to change his body position, and has a remarkable grasping power with his hands. He performs rocking motions and sucking gestures, plays with his tongue, and puts anything in his mouth. Later, when the eyes focus there is an effort to coordinate the random movements in order to grasp whatever object is seen. His next development is in three phases: learning to propel himself with the aid of his limbs; using the forearms to hold onto objects in order to learn to stand and walk, at first unsteadily; and finally mastering and developing greater and greater skill in walking, running, and coordinating himself in eating, dressing, and the other functions of the physically independent human being.

From Seven to Fourteen. Generally their movements are characterized by a nervous, restless energy that requires constant movement and bodily adjustment. Movements of the arms are usually broad, free, and away from the body. Walking with an erect carriage, they tend to appear as if bouncing on the balls of the feet. Their stride is long and the arms swing in free opposition to the leg movements. There is a tendency for the lower legs to turn in slightly. When walking they often break into free turns, hops, or jumps that express their joy in living. When standing they may toe the ground or explore it with their feet. When sitting they usually throw themselves down with their arms moving freely away from the body. They often slouch in a chair and spread themselves out or curl their arms or legs around it.

At the onset of puberty (usually at twelve to fourteen), there are physical and psychological changes which may affect movement, attitude, and behavior. The child must adapt to rapid changes in height and weight. One part of the body may grow faster than others. At this point adolescents may appear awkward, as their skills in using muscles catch up with their rapid growth. They may appear shy or socially awkward as they learn to adjust to the new demands placed upon them by biological changes, social changes, and sexual taboos.

From Fourteen to Fifty. The period from fourteen to fifty is less easily characterized by specific movement patterns. In general, the younger a person is, the freer, quicker, and more spontaneous his movements are.

124

Muscular skill is at its height in the early adult years. The loss of muscular skill is slow and may be offset by the accumulated experience of living. Nevertheless, as age increases movements tend to become less quick, less mobile, and more restricted. The muscles and ligaments become less flexible, the bones brittle, and the movement at the joints stiff. In middle age the leg action in walking is more from the knees and less from the hips, and the elbows are held closer to the sides of the body. In order to maintain balance when walking, the legs are separated more to the side than in youth. Even when the caloric intake is reduced there is a tendency to gain weight for women in their thirties and men in their forties. (There are, of course, exceptions.) Women generally carry the extra weight in their hips and thighs, while men carry it around the midsection. A weakening of the muscles of the shoulders, upper back, and abdomen causes round shoulders and the "pot belly." The lower back may be subject to pain from the additional strain. As with a pregnant woman, the increased weight and the stiffening joints make it difficult to get in and out of chairs, so that the arms and legs must be used more for support.

In the middle years there may be emotional strains placed on women as they pass the age of child-bearing and adjust to the children moving from home. Men may be subject to emotional strain as they adjust to waning sexual energies and the failure to attain goals. How they individually react to the physical changes and the change of their roles in society will influence their way of moving.

The physical movement characteristics of old age are much more pronounced. With advancing age the movements become slower and are done with greater caution. Movement in the joints of the arms and legs becomes restricted, so that bending of the arms and legs becomes difficult. The wrist and finger joints are stiff so that handling of objects is done slowly and deliberately. Sitting or rising is difficult and requires some use of the arms. The legs are turned out from the knees down and walking is done on the flat foot. There may be a shuffling action of the feet. The stance may become wider and wider to give greater stability, which in turn restricts the length of the stride, or conversely the legs may be held close together, again with small, shuffling steps. Movement at the hip joints and shoulder joints is restricted. Very little if any arm movement is used in walking. Turning to look at something requires a turn of the whole body.

As age advances, the muscles of the spine become weakened and less efficient, which makes standing erect difficult, so that the shoulders become rounded, the head drops forward, and the concave torso tilts forward. Walking becomes unsteady. The discs of the spine become compressed, which causes a decrease in height. The body shrinks and becomes lighter with age. Where in youth the skin was tight it now becomes increasingly loose and wrinkled. The eyes are set deep in their sockets. The skull may be reduced in size. There is a marked reduction in the vertical measurement of the face. The jaw bones shrink, causing a loss of the teeth. When the teeth are lost the cheeks become hollow and the lips thin. As the eyesight begins to fail, a studied look at any object becomes necessary. Damage to the inner ear, where amplification of sound takes place, can cause a loss of hearing. Few persons over sixty can hear the higher tones that were easily heard in youth. There may be a tendency to forget things and later a loss of memory for certain events.

The growing limitations on physical activity, the loss of satisfaction in work, the loneliness of retirement or isolation, and the death of close friends will all have some bearing on the attitudes of the elderly and hence on their way of moving.

The important thing to remember is that *there is no one way of moving at any age.* Some children move with the same stiffness as the old and some old people move with the same kind of freedom as the young. Physical mannerisms that commonly come with age are often altered by temperament. It is thus necessary to consider the whole emotional and physical make-up of the character you are playing.

EXERCISE 153

In order to get a feeling of the characteristic movement of youth and age, review the characteristics of people under the influence of joy and prolonged sorrow. Then sit, stand, walk, move objects, and in general try to move as you did in childhood or might in old age.

EXERCISE 154

In order to imagine how it feels to move as an old person, try to remember various analogous personal experiences. Remember when you had a bad cold: your movements were probably slow, your body contracted when you coughed, and you had difficulty in breathing. Remember times when you were so tired that moving seemed nearly impossible. Remember when your muscles were sore, you felt stiff, and moving was painful and so had to be done carefully and slowly. Remember when your neck and shoulders were sore from tension and raising the arms was painful. Remember how your body felt after soaking in a hot tub of water. Remember how, after sitting for a long time, your body felt stiff when you started to walk again. Try moving the way you did under those circumstances.

EXERCISE 155

In order to recall the sensation of moving like a child, try walking around the room, but every few steps do a skip, a hop, or a dance step of some kind and let your arms swing freely in all directions. Try throwing yourself into a chair and throwing the arms up into the air. When you are in the chair bounce around, move the feet, move the arms, change position, and in general find it impossible to hold still.

EXERCISE 156

a. Write three or four sentences on a piece of paper, using the wrong hand. Note the difficulties you have, the concentration necessary to write, the lack of coordination, the slowness of your writing, and the restricted feeling of the movement. If possible, note the similarities in how the very old and very young write. The next time you have a meal, try and study your movements.

b. Fill a pan of water to the very top. Keeping your elbows against the sides of your body, try to carry it across the room as fast as you can without spilling any water. Pour out a little of the water and repeat the walk again. Keep repeating the exercise with less water until you can move across the room comfortably. Now try walking the same way without the pan of water.

SEXUAL DIFFERENCES

Sexual differences traditionally have played a role in movement differentiation. In some cultures the difference in movement is based on actual anatomical and physiological difference between the sexes, while in others the differences are cultural, and in still others there are few differences. Certainly in our own society since some cultural restrictions have been lifted women have become more dominant and outgoing in movement.

Although there are many exceptions in reality, women have traditionally been thought of as the weaker and more passive of the sexes. For men basic movement patterns have been robust and outgoing, but for women, subdued and inner-directed. Ideal women's movements have been characterized as dainty, restrained, and graceful. Gestures were made inward toward the body. Steps were small. Women were to be calm, even demure. Their head or eyes were lowered and their postures were drawn into the body. Ideal men's movements were thought of as bold, large, aggressive, exuberant. Their gestures were made away from the body. Their strides were long. Their gaze was firm with the head lifted high. Their postures were open and spread out in space. Many of these concepts continued in full force until the early twentieth century when the movement toward the emancipation of women became active.

Any characterization based on sexual differentiation must take into account other factors than the gender alone. The traditional viewpoint presented should be thought of as a base of departure for further observation.

EXERCISE 157

In our own society the traditional sexual stereotypes are rapidly changing. To get some idea of the current power of the several stereotypes, observe as many people of both sexes as possible and try to classify them according to the kind of image they project, whether it conforms to the stereotype or not. Try to observe athletes, very socially oriented people, people deeply involved in a cause, leaders, celebrities, and as many other types as possible. Observe their movements and determine to what extent the way they move is a result of their anatomy, their occupation, or their cultural background. Obviously, whether male or female, no two people within a group move exactly alike. Observing like groups of people is only a point of departure for more careful observation.

EXERCISE 158

The next time you watch a sports event, such as a professional baseball game, notice mannerisms of the players that may be theatrical but are not necessary to the game. Then watch young boys playing the same game and notice how many of these mannerisms they imitate. In the same way observe little girls playing house or helping their mothers.

EXERCISE 159

Notice particularly how stereotyped sexuality is used to attract your eyes to the product being sold. These kinds of postures and movements tend to reinforce the stereotyped images we associate with the sexes. Look at advertisements in magazines, newspapers, and on television. Notice how certain postures or movements imply that the user of the product will become either more masculine or more feminine, or will succeed in business, in love, in health, or in being youthful.

LEARNED MOVEMENT PATTERNS

The cultural experiences and social activities which people have habitually participated in are reflected in their movements. The more cultured and sophisticated often economize and slow down in their movements and tend to be less expansive with their gestures. People who are more spontaneous or physically more active by habit and attitude tend to express themselves in freer patterns.

Family background is often revealed through movement, as children tend to carry into adulthood many of the gestures learned from their parents. To understand this influence, try to recall habits and customs of your own that are based on family traditions and practices. Notice also how often children mimic the actions and attitudes of older people.

OCCUPATION

When studying movement, you should take a person's occupation into account. A person engaged in physical labor will generally move with greater freedom than someone in a more sedentary occupation. A person who sits at a desk all day long tends to move with greater restraint than a physically active person. Unused muscles become weak and flabby. Stooped shoulders, sagging posture, restricted movement, and "middle-aged spread" even in the young may be the signs of a sedentary occupation.

The movement patterns required by one's occupation become habitual after a time, so that among fellow workers there are likely to be some similarities not only in talking and thinking, but in movement and gesture as well. People are, of course, individuals and act and move accordingly. However, in observing such occupational groups as housewives, professors, construction workers, hairdressers, athletes, newscasters, sales clerks, salesmen, politicians, and ministers, you should take note of what things they generally share, dress styles, hair styles, ways of expressing themselves, and movement patterns.

NATIONALITY AND CLIMATE

Another important factor in developing a movement pattern is the character's nationality, and, beyond that, the region of his birth and upbringing. In a sense, national and regional differences in movement are the equivalent of accents, dialects, and languages. Atmosphere influences movement. For example, a city dweller, surrounded by buildings and used to crowds and concrete, will move differently from a farmer, accustomed to open space and dirt paths.

Climate also influences movement patterns. Clothing also influences movement. In the heat, where few clothes are worn, movements may be

129

freer and more languid. In cold climates, people move according to the restrictions that their clothing and the climate place on them. Repeated often enough, movement patterns become habitual. For example, those of tropical countries are thought of as fluid and graceful, and those of the Teutonic and Anglo-Saxon countries as heavier, slower, and more controlled.

How people view time also influences their movement. To some people time consists of hours and minutes. It is a precious commodity which must not be wasted: to busy oneself with useful pursuits is virtuous, and to keep someone waiting is rude. To be at work on time means more money for the company. To be late at the bus stop means waiting for the next bus. To get a degree may mean graduating in three years instead of four. Anyone who acts on these assumptions will probably move at a faster pace and be more subject to anxieties and physical tensions.

Other people regard time as a continuous flow without artificial divisions. At most, time for them may be divided by night and day, the changing seasons, or births, deaths, and important family events. As a result, their movements may be more leisurely, their attitude more relaxed, and their relationships with other people less strained.

People rarely consider how their native culture influences their movement patterns. Further, they may regard their local habits as "natural" and others as peculiar. For example, rural and urban people of the same country often regard each other with suspicion, or as quaint or strange. In the same way visitors to foreign countries are sometimes treated badly because they have unknowingly violated local customs. Visitors to "primitive" areas of the world are sometimes amused, shocked, or appalled by customs considered quite natural by those who practice them.

When developing a characterization, consider how customs and beliefs would influence your character's movement. For example, you might ask the following questions.

1. Is the area agricultural, industrial, or both? What is the economic standard of living?
2. Do the parents and children live separated from the grandparents? How are the old treated? Are they revered for their wisdom or closed off from society? Are the children indulged or restricted?
3. Where do the people live: in huts, in separate houses, in town houses, or apartments?
4. How are people educated: by the family, by the tribe, or by formal schools?
5. How do the people entertain themselves? What kind of games do they play? What kind of recreational activities are available?

130

6. Are the laws written for the benefit of the people or for the benefit of the state? Are the laws respected and obeyed or something to be manipulated or tolerated? Are the laws written down and handled by a court structure or tribal council, or are the laws unwritten but enforced by the threat of rejection from society?
7. How do the people defend themselves: with sticks, spears, arrows, swords, guns, planes, bombs, or something else? Are they aggressive and military-minded or defenseless and peaceful?

In any country there are, of course, many different movement patterns, but these kinds of questions can help you to better understand how a particular character might move, based on his national or local background.

The following exercises should be done with the idea of physically exploring a situation as if you had never experienced it before. In each case note very carefully how your body moves in response to the situation. The exercises will have to be altered to fit your locale and the kind of activities available to you.

EXERCISE 160 *Urban Versus Rural Patterns of Movement*

a. Walk at different times of the year in forests, parks, open spaces, narrow streets, crowded streets, streets with high buildings, or any place where you feel the contrast between physical confinement and freedom.

b. Walk on dirt, in mud, on wet and dry grass, on sand, and on wet and dry pavements. Take careful note of how your feet feel on the different surfaces and how your body responds.

EXERCISE 161 *Climate*

a. If the climate where you live changes, take note of your movements on a hot day, a cold day, a rainy day, and a windy day. Visit an air-conditioned place and a steam room. Take a cold shower or a hot tub bath, or swim in a cold and then a heated pool. Place ice on your forehead and wrists. Place hot towels on your face.

b. In order to feel the effects of various kinds of clothing on your movement, first put on loose clothing and move about the room. Next put on three or four pairs of stockings; several pairs of slacks or trousers; several shirts or blouses, sweaters, and coats or jackets; gloves or stockings on the hands; a hat or scarf on the head and over the ears; and scarves around the neck. Move about the room and observe the ways in which your movements are confined.

131

Anyone who has had the experience of moving to a new state or city or even a new neighborhood knows the difficulty of adjusting to new surroundings. The whole pattern of life has to be altered to fit the new area: new streets, new stores, new neighbors, and different neighborhood customs. In a similar way movement patterns and the tempo of a person's movement are affected when he is on a vacation. Movements that were routine suddenly must be changed to adjust to the changed situation.

a. Recall an experience that required such adjustments. Try to remember how it changed your regular patterns of movement. For example, if you moved to a new house it took time to get used to where things were and to go to the right drawer or cupboard without thinking about it. Even rearranging the furniture in a room requires readjustment of your movement patterns until you get used to the new arrangement. To experience this change, move your toilet articles, clothing, books, or writing materials to a new location and observe how the change alters your thinking and movements. You might also rearrange the furniture.

b. Whenever possible, watch filmed documentaries and news films that show people in different parts of the world and observe how they dress, how free or restrained their movements are, the kinds of activities in which they are engaged, their reaction to the weather, and their expressive gestures.

It is the fleeting and scarcely tangible expressions of gestures that unmask a man.

CHARLOTTE WOLFF

6 Developing the Role Through Movement

The ideas presented in this book are only a means to an end. They are tools that can be used to analyze and establish a fully realized movement characterization. The purpose of these techniques is to help you to free your own creativity, spontaneity, and imagination so that you can rise above merely playing yourself and can "become the character." Who you are as a person, of course, will also be in the character. Unless you are able to motivate everything you do from within yourself on the basis of a lifetime of living and observation, the character may not become a believable human being. No matter how much you plan or practice moving as the character would, it is only what you can bring of yourself, your sensitivity, and your knowledge of your craft that can bring a character to life. Directors, teachers, and fellow performers may be of help in the

133

preparation of a role, but the ultimate responsibility for interpretation belongs to the performer himself. In a performance you are, in one sense, very much alone.

PREPARING FOR THE ROLE

There are many different ways to begin preparing for a role. What works for one person may not work for another. Through trial and error you will find your own best method. Nevertheless, at some point you must get a clear idea of the overall production and how your performance fits in with it. This may be easier if you are working with a completed work, in which case you can read it and discuss it as a whole with the director. An overall view may be more difficult when the work is new and still being completed during the rehearsal period. In any event, your creativity and ability to respond quickly to new ideas will be appreciated.

TEXTUAL ANALYSIS

The following ideas for textual analysis primarily assume that you are reading the completed manuscript of a play. However, a similar analysis can be done of an abstract or story ballet, an opera, or indeed almost any kind of theatrical event. Such analysis is also of primary help in understanding your role in a work in progress.

Before deciding on specific elements of characterization and body movements, study and analyze the text thoroughly. You might follow certain steps in such analysis:

1. Read and study the entire play.
2. Decide whether the play can be classed as comic, fantastic, naturalistic, poetic, farcical, satirical, tragic, or historical.
3. Describe in a few sentences the basic premise or theme of the play. For example, "This is a play about a medieval king who, fatally blind to the ambition and greed of his court and to his responsibility as a leader, causes the destruction of his country."
4. Next you should list on paper or mentally note various aspects of the play, such as the kinds of characters in it, the events that take place, and the general mood of the play.

Some of the questions to ask yourself when analyzing character are: What is the outstanding strength or weakness of each character? What is the primary motivating force of each character? What is the state of each character psychologically, physically, and morally at the beginning of the play and how and why has this changed by the end? Has the

134

change taken place because of the character himself, because of another character, or because of outside forces? What is the function of each character in the play? Does each character represent some symbolic force in the play, such as good, evil, power, the rich, the poor, strength, or indecisiveness?

In a summary of events in the play you might list the state of affairs at the beginning of the play, the events that occur later, their causes, and the ways in which they are developed and resolved. This could be done by act and scene or more generally, depending on the structure of the play. Other elements to be analyzed are the general atmosphere of the play and the prevailing mood of each character, each scene, and each act. You might wish to note also any special requirements in scenery, furniture, props, clothing, or accessories. Knowing what physical materials you may be working with, you can anticipate problems and develop ideas for using props and accessories effectively when the time comes.

This kind of preliminary textual analysis is valuable to an understanding of the play and the place of your role in it. Now you should read the play again and begin to look for clues that will suggest how you should portray the character. Study both what the character says and does and how others in the play relate to the character.

What the character thinks of himself and others, how he responds to the events in the play, and what he says and his manner of expressing it all have a direct bearing on how the character will move. As you are studying the text, ask yourself a number of questions about the character that may suggest a physical portrayal. For example, how does the character reveal himself verbally? Does he speak a lot or very little? Does he speak in long wordy sentences or short, stacatto sentences that go directly to the point? Does the character mean what he says or is he saying one thing while thinking and feeling something else? Are his speeches positive or negative in content? Does the character's dialogue indicate that his thoughts are poetic, philosophical, gross, or idealistic, or do they have some other special quality? Is his choice of words significant? For example, does he use words that suggest violence, strong emotions, passivity, or any other special characteristic? Can you find adjectives to describe his mood, his thoughts, his physical condition, or his physical actions?

It is also of great importance to consider how the character relates to the events and other characters in the play. As you read, you might ask a number of questions. What is the attitude of the character toward the others in a given scene? What kinds of relationships are built between the various characters psychologically, socially, and by birth or marriage? In a given scene, where has the character come from and for what purpose? Does he accomplish his purpose? Where is he going and why? How does the character respond physically and socially to others in the scene, to himself, and to the events taking place in the scene? Do the other characters in the play describe the character in any way that might indicate his motives, actions, state of mind, or appearance? How does the character react to what others say to him or about him, and what they do to him? How do the others react to what the character says and does? How much influence does the character have over the events taking place in the play?

CHARACTER TYPES

After thoroughly reading the text you should have some ideas on the type of character you will portray and on how you will play the role. Before memorizing the dialogue or beginning to move about it may be helpful to give more thought to exactly who the character is, what makes him unique, and also what qualities make him representative of all mankind. Historically the theatrical form may change, the playwrights, composers, choreographers, directors, and performers themselves may change, but human nature does not. Certain two-dimensional stock characters or basic human types, for example, are recognizable in many theatrical productions. On the other hand, in some periods the playwright has been challenged to make his characters as complete and realistic as he can. In other cases, playwrights have tried to present "basic universal types" and have to that end imbued them with the mannerisms of stock characters. Composers and choreographers for similar reasons often make a choice between realistic and symbolic characters to represent or embody some particular idea.

The more you know about the play and your role, the better chance you have of bringing to life a character who transcends your own personal way of thinking, feeling, and moving. A theatre craftsman is able to make each role unique and does not have to rely on a stock set of gestures, speech mannerisms, and acting mechanisms for the portrayal of character after character. Therefore, for several reasons, it is important that you be able to recognize stock character types. You should avoid portraying your character with the most obvious mannerisms, unless that is your specific intention. You should be able to sense when you are unintentionally slipping into characterization types or "clichés." You should be able to realize your character as both a universal type and yet as a complex human being influenced by his past and present and his view of the future.

Certain recognizable types have become a part of the theatrical tra-

136

dition. Sometimes the most original and complete creations are in actuality brilliant variations on these traditional theatrical figures, some of which are described below.

Typical stock characters are those of the Italian *commedia del'arte*, which started around the middle of the sixteenth century. There were the conventional lovers (*innamorati*), the comic figures of Pantalone (the money-minded Venetian) and the *dottore* (the doctor of law, who always interferred in other people's business), the servants Arlecchino (Harlequin, a stupid servant who becomes smart) and Pedrolino (Pierrot), the coward Brighella, the braggart *capitano*, and Columbina (Columbine, a confidante and mischief-maker).

The Elizabethan, Ben Jonson, designed more complex characters, but they nevertheless often had one or two outstanding characteristics and were often named symbolically for animals. Restoration plays abound with types based on specific human characteristics, as some of their names suggest: Sir Wilfull Witwoud, Foible, Mincing, Mrs. Dainty Fidget, Old Lady Squeamish, Lady Bountiful, Squire Sullen, Lady Brute, and Lady Fanciful.

Among more recent types are those that derive from the American Western: the "good guy" on the side of law and order, the "bad guy," the school-marm, the librarian, the sophisticated lady or Eastern dude whose city ways are different from those of the West, the saloon girl with a heart of gold, and other supporting types. In the melodrama, there are the traditional hero and villain, and often the innocent maiden who must be rescued from her own folly or the threat of some immoral force.

Many television and movie dramas revolve around the same kind of themes and characters as the Old West, but updated to contemporary times. On television the detective, policeman, lawyer or doctor often fight the same kinds of forces in a contemporary setting. Similar themes are found in dramas of ancient Japan, the Middle Ages, and the Roman Empire.

In other plays, roles are written to represent conventional moral qualities: the seven deadly sins, for example, and the seven virtues in medieval morality plays.

If your character can be classed as a certain type it is advantageous to explore it fully so as to avoid playing it in a thoroughly conventional way. Charting a personality profile for the character may help.

PERSONALITY PROFILE OF A CHARACTER

In order to create a distinct personality you must decide on the basis of the text and your own perceptions just how your character would behave and why. Among the factors to be considered are the motives of the character, the kinds of conflicts he encounters, and the peculiar behavioral patterns the character uses.

In life we are all beset by goals: marriage and a family, success in business, or artistic recognition, for example. Our methods of achievement vary, of course, with our commitment, abilities, and personal circumstances. Characters on the stage are generally similar in this respect to real people, no matter what the style or approach of the playwright. Commonly the conflicts are selected and heightened for dramatic and illustrative effect. Since this is the case, an analysis of your character's personality traits and circumstances should be very helpful in developing a convincing and interesting performance. Some of the questions you may wish to ask in such an analysis follow.

MOTIVES

What is the primary driving force behind the character: that is, what does he want? What is he willing to do to get it, and what, in fact, *does* he do? If he gets it, what happens to him? If he does not get it what then?

CONFLICTS WITH OUTSIDE FORCES

What forces stand between the character and his goals? Do the desires of another character conflict with those of your character? Is their a conflict of wills? Are there obstructive social forces? Is your character in conflict with "fate," or "nature"? Which is the most influential or the most apparent of these forces?

CONFLICTS WITHIN CHARACTER

Are the fears of the character based on real threats or only imagined ones? Is there some weakness in the character that assures his defeat? Does the character have a clear view of himself and his situation or is his self-evaluation faulty in some way? Is the character fully committed to his goals or insufficiently committed? How effective or ineffective is the character in coping with his stresses and anxieties?

BEHAVIORAL AND PERSONALITY DISTURBANCES

Does the character respond to events or other characters in an inappropriate or excessive manner? Does he have control of his emotions? What kind of judgment does he display? Is his thinking ruled by phobias, obsessions, delusions, hallucinations, or superstitions? Does he develop enduring and satisfactory relationships with others? Is the character self-reliant or helpless, clinging, or indecisive? Is the character productive and optimistic or is he pessimistic, stubborn, and inefficient? Does the character act out his frustrations and hostilities on others? Is the character preoccupied or obsessed by something such as drugs, alcohol, food, health, sex, power, pride, envy, money, religion, or virtue? How conventional or unconventional is his behavior, given his historical period and social status?

THEMATIC FUNCTION OF CHARACTER

Does the character represent in some way the universal concerns of mankind by what he believes in, what he does, or what happens to him? Among such themes, for example, are death, success, justice, fate, personal responsibility, honor, beauty, morality, pride, and love.

CHARTING A MOVEMENT PROFILE

No matter what kind of director you are working with, at some point you yourself must develop a movement pattern that characterizes the role you are to perform. To develop a valid movement characterization, you must experiment with movement patterns based on your character's physical appearance, physical condition, and emotional, social, psychological, and dynamic conditioning.

WHAT DOES THE CHARACTER LOOK LIKE?

Without, for the moment, considering the characteristics of your own body, decide on the body image appropriate for the character. Through your body carriage you can project the desired image. How tall is the character? How much does he weigh? Are his torso, arms, and legs short or long? Are they fat, thin, well-muscled, or average? What is the general age of the character? Is he healthy or unhealthy? Is he physically strong or weak? Is his occupation or physical activity reflected in his body condition? Is he attractive or unattractive? Does he like or dislike his body or clothing?

THE MOOD OF THE POSTURE

In a sense, a character is imprisoned in his own body. What he feels at

139

any given time is expressed in some way through his own body, and what he has felt year after year should be stamped indelibly into his face, his posture, his gestures, and his way of moving. To determine these pervasive feelings, you must ask yourself further questions. Is the character's attitude basically optimistic or pessimistic? Is there a single basic mood or emotion that pervades everything the character thinks and does? Does he freely express what he feels or are his emotions repressed? Try to determine effect of such a persistent state of mind on the character's body and movements. For example, will the character stand tall or will he hunch over? Will the character spread his body out or shrink into himself? Will his face be pinched and drawn downward, or relaxed, or drawn upward? Will his movements be free and easy or tight and restricted? Will a particular area of the body, such as the torso and shoulder region, be constricted or held tightly as a result of inner tensions?

THE EXPRESSIVE MOOD OF THE BODY

Try now to describe in general the movements of your character. For example, would his movements suggest that he is cheerful or depressed; bold and adventurous or meticulous and cautious; poised and at ease or fearful, worried, anxious, or agitated; full of trust or suspicious; serious or frivolous; cooperative or uncooperative; or ambitious, jealous, vain, or perhaps feeling shame or guilt?

THE ATTITUDE OF THE BODY SOCIALLY

The character may move in a different way when he is alone than when he is with people. How familiar and comfortable he is in his surroundings and with the situation and the people may be reflected in his movements. What is his position in society? Will he behave according to social expectation? Is he younger, older, or the same age as those present? Is he a leader or a follower? Is he in command or in a defensive position? Is he aggressive or submissive? Is he talkative and engaged in the proceedings or withdrawn and introspective? Is the character frank or secretive? Does he express his own ideas or repeat what others have said? Try to determine in a general way how the character would move if in charge of the situation, confident of his position, and open in expression of his viewpoint. By contrast, try to determine the movement pattern of a person closed in on himself as the result of an inferior position, a closed mind, or limited interests.

LEARNING TO MOVE AS THE CHARACTER

The next stage of developing a movement characterization is to explore it by moving in different ways as the character might move. At this point there should be no attempt to formalize the movement pattern or to tie it to the dialogue or specific blocking. Try only to establish a kinesthetic sense of the posture and movement patterns the character might use. Your aim at this point should be to develop a movement style that expresses his basic personality. How the character walks, stands, sits, gestures, handles props, and relates to himself and others should ideally, in the end, be consistent in style.

There are two basic approaches. The first is to concentrate at the beginning on the way in which the character would use only one part of the body and later the whole body. The second is to work on the whole body from the beginning, ignoring special concentration on individual parts. We will deal with the first method first and the second afterward, in a discussion of walking, sitting, standing, and handling objects. In the beginning exaggerate everything and later, if need be, simplify.

WALKING

One way of beginning a movement exploration is to walk like the character. Isolate each part of the body and decide how the character would hold or move that part of the body.

1. Decide how the character will step onto his foot and use his foot to push off into the next step.
2. Decide the width of the character's stride and how far apart the legs are.
3. Decide how much action there should be in the hip and knee sockets and in the shoulder and elbow sockets.
4. Decide how the character should hold or use his torso and head.
5. Practice walking around the room, moving the part of the body that you are concentrating on until the movement feels somewhat natural to you. Try this with each isolated part of the body, not concerning yourself with how the rest of the body is moving.
6. Try walking while combining some or all of the isolated movements. Continue until they feel somewhat natural.
7. Try to decide at this point if this kind of walk truly expresses the character. If not, experiment until you find a way of moving that does not feel awkward to you and yet expresses what you feel about the character.

141

If you are using the second approach—experimenting with the whole body at once—try the following steps.

1. Walk as tall or as short as you think the character should appear. If you are walking tall, will the body be lengthened vertically upward? If short, will it be contracted downward? Centering the weight on the balls of the feet may give the illusion of added height. Centering the weight on the whole foot can make you appear shorter.

2. Walk according to the weight of the character. If he is heavy, your weight might be pulled downward on every step and centered over the whole foot. If he is light, your body might be lifted upward as if floating, with the weight centered forward over the foot.

3. Decide how spread out in space the body is (forward, backward and to the sides) or how closed in on itself it is. Letting the arms float a little way from the body and holding the head lightly erect may help give the illusion of spreading out. Pulling the head down into the shoulders and pressing the arms against the body may give a "closed-in" appearance.

4. Walk according to the predominant mood or personality characteristics of the character. Can this best be expressed by spreading out and moving freely through space or by holding the body in and restricting the movements?

5. Assign a specific emotional state to the character such as joy, sadness, anger, or fear, and express through walking some varying degrees of emotional intensity. Explore a number of possibilities. For example, if the character is basically an angry type, does he commonly give vent to his anger through aggressive, blustering, threatening body carriage or does he hold his body in check, so that his anger boils beneath a quiet exterior?

6. Assign a single descriptive word to the character, such as prudish, snobbish, puritanical, miserly, lethargic, listless, languid, animated, vigorous, vital, crisp, flustered, apprehensive, high-strung, waspish, grouchy, crabby, gruff, or ill-humored, and explore walking in a way suggested by the word. Is the walk restricted or bold and expansive?

7. Move according to the character's mental state. Basically does the character move forward with ease or as if fighting some resistance? Using imagery may help you to define the quality you want. For example, does the character move forward like an astronaut in the weightless state? Does the character move forward as if walking up to the neck in calm waters? As if walking hip-deep up-

stream in a river? Like a horse at a loping gait, a trot, or a gallop? Like an enraged bull on the charge?

8. Try to determine the character's attitude toward time and how this would influence his walk or walking. As a rule does he feel that time is limited or unlimited? How does this affect his speed and rhythm? Does the character dawdle, creep, saunter, plod, trudge, shuffle, drift, stroll, glide, float, stalk, amble, pace, march, lunge, pursue or dash? Does the attitude and the rhythm of the walk express anything about how the character views himself? For example could it be classed as strutting, swaggering, jaunty, haughty, aloof, buoyant, ethereal, abandoned, giddy, restrained, or inhibited?

9. The practical purpose of a walk will influence the walk to some extent. Set up a series of goals that the character wishes to attain or a series of specific mental problems and try to express them in walking as the character would. Try to combine them with ways of walking which you have already explored so as to develop a basic walking style that remains even though the purpose of the walk changes. You might walk toward the library to return a book or away from it after returning the book; into a strange restaurant to find a table; up to the cashier to pay the bill; out of the restaurant after the meal; or to a bookshelf to select a book and away from the bookshelf with the book. You might walk into an employer's office to ask for a raise; into a darkened movie house; into a dark room to look for the light switch; into a room full of strangers; into a room full of friends; into a classroom at the beginning of the school year; or into the schoolroom at the end of the year. Some examples of mental states that will alter the walk are: agitation (pacing the floor), impatience (waiting for someone who is late), concentration (trying to decide what color to paint a room), anticipation (going to answer a knock at the door), and curiosity (going to a window to see who is walking down the street).

SITTING

After the walking exercise, try the same kind of movement definition in sitting in and rising from a chair and later a sofa, a stool, the ground, and other locations. Try different ways of getting into and out of the chair, and assume different postures while seated. How would the character approach the chair, for example, if he had been invited to sit down, were tired, were in a hurry to leave, were uncomfortable in the situation, were in control of the situation, and so forth? How would the character approach the chair if it were at a table, at a desk, against the wall, in the center of the room, or in some other special location? How would the character then get into the chair? Would he use his arms? Would he face toward or away from the chair before sitting? Before lowering his body to the chair would he bend his torso toward the chair, or forward away from it, or would his torso remain upright? Once he were seated, would his hips be well back in the chair, in the center, or on the edge of it? Would the torso be slouched forward, leaning back, upright, or twisted? Would the knees be together or apart? Where would the legs, arms, and head be placed? Where would the hands and feet be placed? Experiment with motives and movements in the same way when rising from a chair. Once standing, would the character move directly away from the chair, pause briefly, or remain standing by the chair?

STANDING

While standing, explore poses the character might assume if he were waiting, talking, listening, thinking, under emotional stress, or in a given state of mind. Assume stances that would reveal various degrees of interest in the proceedings at hand. Would the weight and the body be shifted forward or backward or be in equilibrium? Experiment with various positions in which the character might place his feet, arms, torso, and head. How would he position his body if facing the imagined action, facing away from it, or standing sideways? Experiment with different postures and relationships to space and objects when the character is standing against a wall, in the center of the room, or next to something such as a chair or table. For example, if the character were next to a chair, could he use the chair to add variety to his postures or reveal something about his state of mind or attitude? He might simply stand near the chair, or he might place one or both hands on the chair, place a foot or a knee on it, or lean on it, for example.

HANDLING OBJECTS

Another way of developing characteristic movement patterns is to find ways of using or relating to objects. As before, it is unimportant whether the character actually uses these particular objects in the play. Later you may want to explore the handling of objects belonging to a certain historical period.

144

Determine how your character, given his basic personality, might be expected to take hold of a particular object. Might the action be described as seizing, grabbing, snatching, grasping, gripping, clutching, clenching, or slinging, or perhaps cradling, fondling, caressing, or enfolding? The character's attitude to the object will affect energy and rhythms used by the character in movement. For example, to seize something requires greater force and speed, while to grip something requires greater pressure or tension. Such actions might indicate an aggressive and domineering nature, or entirely different qualities, depending on the circumstances.

Pick various real or imagined objects up, handle them, and use them as your character might. Vary the sizes, shape, weight, and texture of the objects and pick them up from various locations: from a table while standing or seated, from the floor, and from a shelf overhead. Among objects you might pick up are a piece of paper, a notebook, a magazine, and a newspaper; a ring, a watch, a necklace, and an earring; a teacup, a demitasse, a mug, and a cup and saucer; a brandy snifter, a liqueur glass, a champagne glass, and a tumbler; a marble, a golf ball, a tennis ball, and a basketball; a book of matches, a pocket lighter, and a table lighter; a bar of soap, a carton of butter, a carton of laundry detergent, a full, half-full, and empty milk carton, a carton of cokes, a can of olives, and a can of coffee; a wash cloth and hand towel; a package of potato chips and a package of marshmallows; a top hat, derby, cap, and bonnet; and a ball point pen, a cane, a sword, and an umbrella.

Up to this point your exploratory actions have been somewhat confined physically. In order to explore the space around you more fully, devise actions that require a broader use of the body in space. Among such actions are pitching a baseball, throwing a basketball, rolling a bowling ball, throwing salt over your shoulder, passing a note to someone directly behind you, sawing a board in half, screwing a light bulb into a socket directly over your head, tossing a ball up into the air, swinging a golf club, batting a ball, chopping blocks of wood, pounding a nail into the wall, pulling a rope in a tug of war, pushing a shopping cart up a hill, lifting and lowering barbells, and picking up buckets of water and passing them to the person next to you in assembly-line fashion.

Each of these actions requires a different area of space and a different amount of energy. As you move through these or similar actions be aware of the pathway of the movement through space and the amount and kind of energy appropriate for the action. First do the action as you yourself would do it and then the way the character might do it. Try to determine whether the character would habitually make gestures using any of these kinds of energy or spatial pathways.

145

Invent a scene for your character that will include walking, sitting, standing, and using real or imaginary props. In playing the scene, try to keep a stylistic unity of movement based on how you feel the character should be portrayed. In such a scene, your character could, for example, enter a friend's room where there are people present. (Given his basic nature, will he pause at the door or proceed directly into the room?) Once inside he is introduced to a woman and a man of his own age, an older woman and an older man, and a young child. He next excuses himself and exits through a door which he has to open. He goes to a desk, pulls out the chair and sits, takes out paper and pen, and writes a short note. (If the character has been bold, aggressive, or confident in his movements, will his handwriting be large and bold? If the character has been shy, quiet, or secretive, will his handwriting be small and neat?) At this point the character accidentally knocks a letter opener behind the desk. He goes behind the desk, picks up the letter opener, returns it to the desk, and sits back down. His friend then enters through a door directly behind him, carrying a small tray. The character turns and greets him. The friend sets the tray on the desk. The tray contains a teacup, a teaspoon, a pot of tea, a pitcher of cream, and cubes of sugar. The character fixes himself a cup of tea, which he drinks.

A second approach to developing movement phrases is to select a number of characteristic postures, gestures, or walks the character might use and move from one to the other. The actions do not have to be related by a theme, although that can help the movement to flow from posture to posture more easily. The idea, as before, is to get into the body the feeling of moving in ways typical of the character. Among the motivations for postures are thinking, listening, watching, preparing to speak, and not being involved in the proceedings. The first time, move from one posture to the other *without* planning your moves or postures. Then repeat each posture individually, analyzing the carriage of the torso and head and the positions of the legs, feet, arms, and hands. Try to determine if the posture really expresses what you want it to. Is each posture significantly different from the others and in itself visually interesting? Alter the postures accordingly. Then in slow motion move from one posture to the other several times without stopping. Keep repeating the sequence identically from start to finish like a round, and on each repeat speed up the action. At this point it should be like a dance sequence or like slow-motion tai chi ch'uan, in which the action never ceases.

Once the movement is under control, feels somewhat natural, and can be repeated exactly the same way each time, you may want to set up a simple theme involving several postures. For example, you are in the midst of thinking when two people enter the room. They are discussing something and do not see you. Not wanting to stare at them you only listen to what is being said, but the conversation is such that you feel compelled to watch them. You feel that you should warn them of your presence, but you decide that you do not want to embarass them, so you withdraw. When you repeat this movement phrase, take as much time to hold a posture or to move from posture to posture as is needed to make the expressive intention clear.

Now develop some themes for movement phrases involving walking across a room. Be sure to establish first the basic posture and walk that you have decided are appropriate for the character. For your movement phrase, you may use the same five ideas: thinking, listening, watching, preparing to speak, and remaining uninvolved. In any event, try to establish each idea in a number of steps: perhaps four. For example, enter the room thinking in four steps, hear someone talking in four steps, look up and watch them in four steps, move toward them ready to speak in four steps, and decide against it and pass them by in four steps.

You can also use a prop for this exercise. For example, you might be seated on a park bench reading a book, feeding the pigeons, or eating your lunch when people enter and sit beside you. While standing you might be cleaning your reading glasses or wiping your brow with a handkerchief because of the heat when two young children enter and proceed to tramp through the nearest flower bed.

ANIMAL IMAGERY

Another approach to characterization is to use animal imagery. While reading your playscript, think of an animal that might behave in a manner similar to that of your character. List words and phrases descriptive of that animal and how it moves. Some common phrases are: sly as a fox, slippery as an eel, sings like a bird, and struts like a bantam rooster. If you have chosen a hen for your animal, you might, for example, include on your list such phrases and words as: domestic, quick and jerky movement, considered dumb, subservient to an aggressive male, produces eggs, eats pellets of food, has a long beak that can only open and close, has beady eyes, has a long neck that is thrust forward from the main part of the body, has a rather oblong body with a protruding chest, "speaks" with a guttural sound, turns head from side to side to see ahead (since it can only see to the side), movements are primarily quick jerks of the head from side to side, quick pecking motions to the ground when eating,

147

occasionally a flapping of the wings for short flights, and walks on "flat feet." Since the list is only for your use, it can be quite informal.

If you want to use the chicken as a point of departure for gestural and movement patterns, first try to mimic as precisely as possible the movements and apparent attitudes of the chicken, so as to get these new rhythms "in your body." Then you can try to simulate the hen's movements *as a human being might make them.* For example, instead of moving the eyes alone to look around the room, shift the whole head quickly and frequently, and move the arms primarily from the shoulder joint and the legs primarily from the hip joint. Thrust the neck, head and chest forward so that there is a forward slope of the body from the pelvis to the back of the head. In talking, you might drop the head forward repeatedly, as if pecking at food. The aim is to use the animal movements to enrich a human characterization.

COORDINATING MOVEMENT AND DIALOGUE

A director may give you very specific directions on where and when to move on stage and may supply you with motivation for each movement. On the other hand, he may at times ask you to move simply because he wants to establish a balanced grouping or a more interesting stage picture. In some situations the director may function like a traffic cop, giving directions as to where and when to enter, move, and exit, but leaving to you the details of how you move and why.

Whatever the directorial approach, you will probably want to move in character, or, if not, to move out of character for certain well-considered and artistically justifiable reasons. The material below suggests approaches to analyzing, interpreting, and adding variety to your movements and gestures. The ideas suggested are useful whether you work out your movements and gestures privately or are attempting to realize the instructions of the director.

At first review some of the ideas and methods already discussed.

1. Determine what the production intends to accomplish and how each scene contributes to the overall purpose.
2. Determine what the rhythm of the production should be as a whole and scene by scene.
3. Determine how each scene begins, evolves, and builds to a climax.
4. Determine the artistic shape of the production as a whole. It probably has a beginning that sets the ground work for what will follow, a middle that develops the implications of what was set up in the beginning, and a conclusion that resolves in some way what has preceded it.

5. Determine how through your careful selection of gestures and movements you can contribute to the overall plan of the production.
6. Review what a given scene is intended to accomplish, the basic rhythm of the scene, and how the basic rhythm of your character's movements will contribute to the rhythm of the scene.
7. When entering or exiting, know where you have been, where you are going, and why. Decide what your attitude is toward the place and the people present. Plan ways in which your movements will make this clear.
8. Determine the purpose of each of your actions or gestures. Does it reveal something about the character or his attitude, is it intended to further the plot in some way, or is it for a functional purpose: for example, do you move to an area to pick up a prop a few lines later? Many movements serve more than one purpose.
9. If there is a choice of actions, determine which is most effective. For example, would standing, sitting, walking, or a shift of weight be better than a movement of the arm, head, shoulder? Think in terms of the whole scene and production and not only the isolated moment.

In planning movement there are generally five common errors:

1. Remaining immobile throughout the scene.
2. Making random and uncoordinated motions.
3. Making gestures with the elbows held at the sides of the body.
4. Making gestures that are unrelated to the thoughts being expressed.
5. "Acting out" an action, gesture, or the use of a prop rather than performing as you would in real life. The impression of "acting out" usually involves exaggeration and inaccuracy.

In most cases it is desirable that gestures and body movement evolve naturally out of the dialogue, lyrics, or dramatic context. Therefore, the following kind of detailed analysis may not be desirable or necessary. However, it is one method of working so it is presented for those who may find it useful at certain times. For example, it can be particularly useful when setting movements to the lyrics of a song or an aria, which at times may require a theatrical effect rather than a natural one.

Decide which gestures will:

1. Reinforce the dramatic meaning of a line, a lyric, or a movement phrase.

149

2. Help to make clear to the audience some logical point about the plot.
3. Reinforce a psychological point about the character or his reactions to events or others.
4. Be theatrically effective.
5. Be visually attractive.

Decide which gestures are habitual for the character and what other gestures could add effective contrast to the normal movement pattern. Decide which gestures are important and should be emphasized, which are minor and should be understated, and where there should be no movement. Plan when movement pauses would be most effective: before a gesture, after a gesture, during a gesture. Plan where a movement or a gesture should be accompanied by breathing in, breathing out, or holding the breath.

Clarify in your mind the meaning behind each gesture. State in words the meaning of the gesture. Perform the gesture as expressive of the idea behind it. Experiment with a single gesture to see how many different meanings it can suggest. For example, a shrug of the shoulders can convey indifference, doubt, a chilled feeling, ecstasy, and many other feelings. Next, combine movements of the face, the eyes, or some other part of the body with the gesture to make it more effective. Do the same movement, varying the quality of the movement by experimenting with changes in motivation, speed, rhythm, tension, and the use of space.

Determine whether the gesture should be:

1. Incomplete or stopped midway, perhaps to suggest hesitancy or uncertainty.
2. A full, free, and open gesture made away from the body, perhaps to suggest assurance or a concern with something outside of the character.
3. A gesture made toward the body or restricted, to emphasize a personal concern or the character's restricted state of mind.
4. A gesture that is repeated for emphasis.
5. A gesture that uses an isolated part of the body such as the face, head, fingers, hand, lower arm, or foot.
6. A gesture that involves the whole body in some way.

Next plan where each gesture should come in the context of the scene and the production as a whole in order to build to a climax. Decide if the climax can best be achieved by increasing or decreasing the amount of movement.

Ideally, of course, gestures should arise from the dialogue or lyrics. If the movements will be done while singing, plan where the gestures should accompany or punctuate the words. Decide on which word a gesture will effectively strengthen the meaning of a line.

Read a paragraph, phrasing related thoughts together and planning where you can take a breath. Plan places where you can shift the body or make a gesture. Then practice reading the lines for meaning, coordinating your speech patterns and body movements.

Once you have set movements that feel right and comfortable to you, avoid falling in love with any of them. They may be ineffective. Analyze them and throw out those that do not meet the standards you have set for your performance. Keep experimenting, altering, and changing until you arrive at the very best way of presenting the character. However, it should be pointed out that out of courtesy to your fellow performers there is a point where you must let them know what you are going to do so that they can play the scene effectively with you.

As a final reminder, avoid scene-stealing and inappropriate acting pyrotechnics that damage the production as a whole. A discerning audience recognizes this for what it is, although the naive may be impressed. Analyze and intellectualize as you need to before the performance, but in the performance, using the energy and movements appropriate to the character, be spontaneous and free.

7 Moving in Costume and Period

One of the most difficult tasks for many beginning performers, and sometimes for experienced performers as well, is to wear a costume with a sense of naturalness and to project the image that the performer, character, and costume are moving as one in the dramatic context.

Prior to the movie camera there was really no way to know precisely how people of earlier times moved. We can study pictures and artifacts and read the etiquette books that influenced the manners and styles of a period. We can study the period dances, literature, clothing, utensils, and furniture. In effect we can study every aspect of a culture that may give some insight into its movement patterns and still we are left with very little material with which to reconstruct its actual way of moving.

One key to period movement is the constancy of man's basic nature.

152

His mode of living may have changed radically over the centuries, but as a human being with drives, instincts, and appetites, man does not change. Any character portrayed must be brought to life first of all as a human being. Mannerisms may suggest a period, but they must be integrated in a believable performance; conventions of movement in our own age mark us as products of a culture and yet help to define us also as individuals. It is not enough simply to research the common mannerisms of a period and then mechanically graft them onto a performance. Even though characters may be wearing strange clothing and may move in ways that seem remote from our own, they are nevertheless human beings reacting physically and psychologically to their view of reality and the society they represent. The performer must integrate that viewpoint into his performance before it can be accepted as natural or real by an audience.

You must therefore determine what kind of world your character lives in and how it affects his viewpoint and behavior. Is the prevalent viewpoint puritan, hedonistic, god-centered, revolutionary, or something else? Is the government theocratic, monarchic, dictatorial or democratic? Is the society agrarian, feudal, mercantile, industrial, or militaristic? Who are the leaders and who are the workers in the society? What is the prevalent attitude toward foreign countries and ideas? Is the country an island? Is it coastal or inland? How does its location affect its means of defense or trade? How do the people get their information: from newspapers, magazines, a towncrier, gossip at court, letters, or travelers? How do they travel, and on what sorts of roads? What kinds of entertainment are available: tournaments, fairs, strolling minstrels, public dances, court plays? What are the daily living conditions? What about shelter, heat, eating utensils, food, drink, bathing facilities, social customs? Race, religion, distinguishing national characteristics? Such questions should give some insight into your character's outlook on life and how it might influence his movements and general physical behavior.

Very early in the rehearsal period you should have some idea of the kind of costume you will be wearing and how it will change your manner of moving. If from the beginning you know the limitations imposed on you by a costume and how you can move in it to advantage, you will avoid much needless correction later. It is foolish to assume that if you put on a strange costume for the first time at dress rehearsal or on opening night you will suddenly and magically begin to move as the character should move in the costume. By that time the mannerisms, movements, costume restrictions, and blocking should be so drilled into your body and mind that you can move through them naturally without having to give them any undue thought.

153

Although it is not always possible to rehearse in a costume exactly like the one you will wear in performance, you should begin early in the rehearsal period to wear clothing (including shoes) and work with props that are similar in essential ways to those you will perform with.

COSTUME RESEARCH

In preparing for a period role, look at costume books, pictorial magazines, paintings, and museum exhibits for the detail on posture and handling of the costume. Ask yourself the following questions.

1. What sort of costume is being worn, and for what puorpose? What sort of silhouette do the costume and body make?
2. Does the costume conform to the outlines of the human body or does it distort some part of it?
3. Is the costume tight or loose and flowing?
4. Is it perhaps tight on the top and loose and flowing on the bottom, or the reverse? How would movements have to be changed to compensate for this?
5. What is the fabric of the costume and how would it affect movement? Is it light and flowing, like chiffon? Does it cling to the body like silk? Is it heavy and coarse, like some kinds of wool?
6. What kinds of sleeves and collar does it have and how would they affect movement?
7. What are the hair styles of the period? Would they affect carriage?
8. Are hats being worn? How would this alter movement?
9. Is the person in the picture or exhibit holding something? If so, is he holding it in some characteristic way?
10. Is the person wearing jewelry, a knife, a sword or other accessories? If so, how are they worn?
11. How are his head, torso, arms, and legs positioned? Can you identify certain expressive gestures?
12. How are the hands held and where are they placed?
13. What kind of mood or attitude does the posture suggest?
14. If there is furniture in the room, what does it look like, and how are the figures grouped around it?
15. Do you note any expressive gestures with costume? For example, a lady in certain eras might lift the front of a long dress to reveal the tips of the shoes.
16. What is the mood of the picture? Does lighting or coloring suggest anything about the situation or the people?

154

In order to get a kinesthetic feeling for moving the way people might have moved in a particular period, select from pictures a number of characteristic poses, gestures, and actions that characterize the manner and style of the period. Practice these static moments and then combine them in flowing movement patterns. As you go through them, try to imagine yourself in the clothing of the period. Experiment with ways in which you think the character might move given the restrictions of the clothing, the conventions of the period, and the dramatic situation.

CLOTHING: OTHER CONSIDERATIONS

In addition to knowing precisely how clothing will affect your movement, a thorough knowledge of the cultural functions of clothing during various historical periods will help you to devise appropriate movement.

Despite the variety in his apparel, throughout history there has been very little change in the way man puts on his clothes. Simply stated, the main part of the costume may be hung on the body and suspended from the shoulders or waist (in modern style), or the costume may be draped on the body in some way and be suspended from one or both shoulders or from the waist (as, for example, with the Greeks).

Anthropologists, recognizing exceptions, have in general reached consensus that man's dress, besides answering the need for warmth, fulfills one or more of the following cultural needs.

1. Man has decorated his body with ornaments and paints so as to control his environment: to ward off evil spirits, to assure fertility, or to assume a good hunt, for example. Aesthetic effectiveness is in such cases closely related to religious and practical aspirations.
2. Ornaments have been worn to show social or political status.
3. Clothes have been designed to distinguish between the sexes and to emphasize parts of the body considered desirable, attractive, or erotic.
4. Although there are exceptions, clothes in general are lighter in hot climates and heavier in the cold.

Mankind has in various periods decorated his body with objects on the head, the neck, the arms, the waist, the legs, the fingers and even the toes. Women have variously exposed or hidden their breasts. cinched their waists, enclosed themselves in corsets, hidden under voluminous skirts and padding, often abused their bodies in various ways to conform to the prevailing idea of beauty. Men, in keeping with their aggressive role, have often worn clothing that allowed for greater mobility. Social

155

status or rank has usually been indicated in clothing by style, fashion, fabric, and sometimes color.

In the theatre it is helpful to study the costume to be worn in terms of its function, based on the distinctions outlined above. If certain aspects of the costume are richly adorned or prominent, these may be considered as possible areas to "show off" in movements. For example, if you wear a headpiece, jewelry about the neck, jewelry on the fingers, beautiful shoes, or beautiful or interestingly shaped sleeves, you should gesture and move the body so as to display the feature properly. If status is implied, determine how you can use the costume to make that status clear. Is the costume designed to accentuate or conceal certain features of the anatomy? For example women's breasts have been exposed and hidden, but in both cases the intent has often been to draw attention to the breasts. The kind of clothing worn according to the climate will affect the movements. Clothing may suggest climate. Light clothing will make it easier to move, heavy clothing will be more restrictive. Theatre costumes may also suggest in color or fashion the psychological state or mood of the character wearing them. For example, sometimes black suggests mourning, white purity, red passion, and bright colors gaiety.

Following are some of the features to look for in a particular costume and a consideration of how they can affect your movements. In many cases the comments will apply to women's costumes only, but in others they will apply as well to men, who have often worn short or long robes, tights, or loin cloths.

The important thing is to note the way a costume hangs from the body and what its prominent features are. You can start by studying the *silhouette* made by the costume when it is worn (figure 7-1). Does it hang full and loose from the shoulders? Does the garment reach to the waist, the top of the thighs, the knees, the ankles, or the floor? Does it fit tightly on the torso or lower part of the body? Is it belted in at the waist? Is there a cape hanging from the shoulders? How far down does it hang?

If the costume fits tightly at the torso, your movements will be restricted in that area. You will have to stand erect and think of moving the torso as a single unit when turning and bending. If the costume hangs loose from the shoulders, you will have to stand or sit erectly so that the costume does not buckle unattractively, but hangs gracefully in long straight lines or folds. If the legs are visible, then the placement and movements of them will be even more important than when they are hidden under a long robe or dress.

Study the area of the neck and head. Is there a collar? Does it cover the whole neck up to the chin? Does it extend up behind the head? Is the

7-1

Costume Silhouettes

157

head hooded? Is the face framed by a headpiece, a hat, or veils? Is a wig worn, or is the hair piled on top of the head? If the torso covering is cinched in by a tight waistband or corset and the head is bound up by chin straps, a high collar encircling the neck, or a high collar behind the head, movement will be severely restricted and isolated mainly to the arms and legs. In order that the face can be seen, the body will have to be directed more toward the front. Twisting and bending of the head will be restricted. If veils are suspended from large hats or if the hairpiece or hat is large or high, the head must be carried erectly in order to prevent losing the hair or headpiece, and to display them properly. Bowing will present special problems.

Consider the shape, fit, and size of the sleeves. In some periods the sleeves are very tight, so that raising the arms above the shoulders is difficult. At other times the sleeves are full and so wide that the ends reach the floor. The sleeves may be puffed so that the arms have to be carried away from the body. If the sleeves are distinctive, look at pictures and devise ways to hold the arms to show them properly.

If the costume has tights (with or without breeches over the pelvis or thighs) they should be pulled up to cover the body like an outer skin. Accessories, such as ribbons or ruffles for the knees or ankles or elaborate buckles or ribbons for the shoes, should be positioned and worn for maximum decorative effect. Study carefully such common costume accessories as jewelry, ribbons, muffs, handkerchiefs, shoes, gloves, fans, umbrellas, handbags, and belts. Note how they should be handled and worn.

Long dresses present several movement problems. Dresses of certain periods require particular modes of body carriage. The length of a dress, a hoop, a bustle, or a train influences the kind of movement that can be done gracefully while walking, standing, or sitting. As previously stated, the fit of the bodice and the cut of the collar and sleeves also influence movement.

If the dress touches the floor, the actress must lift it when walking so as not to step on it. A common method which will work for most costumes is to place both hands a little forward of the sides of the body and lift the front of the dress a few inches off the floor so as to make a graceful draped fold of material between the hands. This assures the performer that the two feet are clear of the material. One can also lift the skirt in the same way but with only one hand. A third method is to lift the material with one hand but pull it diagonally upward and off to one side. Whatever method is used, attention should be paid to the drape of the material. The material should be lifted only high enough to reveal either the underskirts or shoes, according to the fashion of the period.

In the early twentieth century, long dresses were lifted with only two fingers rather than the whole hand. In the late middle ages, from the front at about waist level, the outer dress might be lifted as high as the abdomen. The hem was approximately one foot off the floor. With the head forward, the torso was bent backward and the abdomen was extended forward to present the ideal of style, the "pregnant look." The body was "S" shaped. High or wide headdresses with silk or linen veils attached trailed over the shoulders, often reaching to the ground.

In certain periods of history, trains of varying length have trailed the floor in the back. A woman standing or moving across the floor with a long train can create a beautiful effect. On stage the train must be manipulated to assure that it drapes gracefully and is not stepped on. This needs early rehearsal for both the actress and the other performers, who must learn to allow for the trailing skirt. Stepping on the train may not only rip the costume but may also destroy the theatrical illusion.

When a woman walks on stage with a train extending out behind her, several things must be considered. When she walks forward there is no danger unless the train gets entangled on the set or someone steps on it. If she is walking forward to a spot on the stage where she will remain stationary, she must consider the final drape of the dress. It can remain behind her, which presents no problem. The other typical line of the train is to have it curled around the legs. In order to achieve this she walks forward, stops, and makes a quarter- or half-turn with the whole or upper body. In order to get out of the position, she can gently move the train to the side with the toe or more gracefully turn back in the original direction, so that the train will pull out straight when she walks. A similar visual line is created when sitting on a chair. The actress faces the chair, turns and sits so that the train gracefully curves around the legs. The train can also be gently kicked in to the desired position. When rising the easier solution is to move the train aside with the hand first. The important things are the effect created by the drape of the material and the position of the train in relation to the next steps to be taken.

Another garment requiring practice is the hooped skirt. Elizabethan skirts, hanging to the floor, were suspended over a single wheel that hung from the waist, sometimes with padding. Later the wheel was made broader at the sides so that the dress was wide rather than round. Much later the frame was extended backward, leading to the bustle. The other common bell-shaped skirt, associated with the early Victorian era, was the hoop skirt. The skirts were worn over circular hoops suspended from the waist. The circumference of the hoop increased as it neared the floor. Pantalettes with lace frills were worn underneath and could be seen when women walked, sat, or lifted their skirts.

159

With these skirts the important thing is to walk from the knees and not the hips, in order to keep the motions easy and flowing, with very little rise and fall in the body on each new step. Resting the hands on the sides of the skirt can help to eliminate excess swaying, although a certain amount may be desirable. If the woman thinks of herself as centered in the skirt while moving smoothly ahead, exececesive bounce of the skirt can be avoided. It should be noted that with the Elizabethan hoop skirt a slight amount of vertical bounce was considered desirable.

Sitting in a hooped skirt takes some practice. When approaching the chair, guide the skirt past the chair in order not to knock it over, then lift the skirt or hoop in back before sitting. Be careful not to sit on the hoops. When rising, step well forward so that the skirt falls clear of the chair.

The supporting framework for the wheel-shaped, lateral-shaped and bell-shaped dresses is enormous. The workmanship and the size of the dress were indicative of social class. The bell-shaped hoop skirt often was six feet in diameter. In the mid-eighteenth century the laterally shaped skirt might be up to eighteen feet across. To add to the problem, enormous headdresses decorated with such things as ships could extend several yards above the head of the wearer. In an ordinary performance one does not usually have to deal with dresses of these dimensions, but they can still be seen on the stages of today in nightclub performances in such places as Las Vegas and Paris.

Skirt size must be taken into account in staging. The performers may have to be widely separated, and men must sometimes approach women at arm's length. It is wise to start rehearsing early in the wire framework of such a skirt so that problems can be solved early.

In the late nineteenth century, the framework was extended out in back to form the bustle. The skirt was gathered up at the bustle and hung to the floor in a train. The body bent forward slightly at the waist because of a tight corset. The posture was known as the "Grecian bend." Walking in the costume is usually no problem, and if the costume does not force the wearer forward, the performer should try to acquire this characteristic bustle stance. Trains were swept aside with a twist of the hips, never with the hand.

A few comments should be made about the hobble skirt. As the name suggests, the performer must take small steps because of the very narrow circumference of the skirt. When sitting or standing, legs must be held close together. Expressive movement must be limited primarily to the upper body.

Although it is outside the scope of this book to deal in the specific mannerisms and conventions of a particular era, certain general customs and widely used stage props will be discussed.

BOWS

A bow is a form of acknowledgement or respect shown to someone else. It may be no more than a simple nod of the head or it may be a complex ritual governed by specific rules. How a character bows may reveal much about his station in life and command of fashionable manners. For example, awkwardness may suggest that he is ill-bred, grace may suggest his good breeding, and affectations may suggest his foppishness.

For women a common period curtsey is to step sideward or forward on one foot, at the same time bending the back leg and placing the ball of the foot on the floor (figure 7-2). The feet are separated, with the weight centered between them. They then descend in a vertical line toward the floor. The bend in the legs may be slight or enough to bend all the way to the floor to a kneeling position. The dress can be held out to the sides or lifted slightly while curtseying. Often the arms are held to the side 'of the skirt or raised to the side toward the level of the shoulders. If hats or wigs are worn, the head must be kept upright.

Men's bows may consist of a simple nod of the head or a bend at the waist. The feet are often together with the knees straight. Period plays require more difficult bows. In the Elizabethan period a common bow was to turn out the legs, stand on the left foot, step back on the right foot and bend it, straightening the left foot in the forward position, at the same time with a flourish removing the hat with the right hand and sweeping it across the body, which is then bending forward (figure 7-3). As the body straightens the weight is taken on the left foot, the right foot is brought forward, and the hat is replaced. The left hand may be at the waist. The Restoration bow is similar to the Elizabethan bow except that the front foot is brought back to the other foot rather than the back foot being brought forward as in the Elizabethan bow. The hand may be

7-2

Curtsey

7-3

Bow

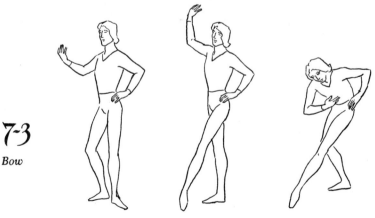

161

brought to the heart at the beginning of the bow and presented with the palm up to the lady during the bow. The hat, carried under the left arm, may be swept behind the body or brought to the heart by the right hand. The left hand may be at the waist or holding the hilt of the sword. If a handkerchief is held in the right hand during the bow, several circular motions of the wrist to twirl the handkerchief precede the bow. Male servants generally bowed from the waist with their heels together.

HANDKERCHIEFS

In the Restoration period men often carried a large lace-trimmed handkerchief in the hand, the cuff, or the outside pocket of the coat. The center of the handkerchief was held by the first and second fingers or the second and third fingers. The handkerchief was allowed to fall over the palm or the back of the hand, which was often raised to show off the elaborate cuffs of the coat. Women sometimes held a handkerchief between the first and second finger, with the arm extended down in front of the skirt.

FANS

Fans can be waved by action originating in the shoulders or wrist. In some periods mirrors were suspended at the waist by a ribbon or concealed in the fan. The mirror in the fan allowed for surreptitious looks over the shoulder or study of one's own face. The fan could be used to shield the face to suggest modesty, or to shield flirting, kissing, or gossiping. It could also be opened or closed or waved vigorously in reaction or for emphasis. The actress may suggest a gentle reproach by tapping someone on the hand. Use of the fan should be discreet and carefully planned or it can become a bothersome distraction.

SNUFF BOXES

Snuff boxes were used ceremoniously in the Restoration period. The box was taken from the waistcoat pocket with a flourish and the top of the box was tapped to dislodge grains of snuff adhering to it. The box was held in the left hand and the snuff was taken out with the thumb and second finger of the right hand and brought to each nostril in turn or it was applied to the back of the left hand and lifted to each nostril. A third way was to raise the curled finger to the right nostril with the palm facing

162

out, and then bring the fingers to the left nostril with the palm facing inward. The box was closed with the left hand and returned to the pocket. The handkerchief was used to flick off any spilled grains from the cuff or shirt front.

SWORDS AND RAPIERS

Swords and rapiers are worn on the left at the waist. When one is bowing, a sword should not be allowed to rise in back. When one is sitting, it should be guided out of the way of the chair. When the jacket has tails of some kind the sword can be used to gently move the tails aside. When one is walking, the sword should not be allowed to swing too freely.

In using stage props, ask yourself a number of questions about them. How can the prop be handled most effectively to reveal something about the character in regard to his social status, his command of the manners of the period, or his state of mind? What technical problems does handling the prop present? What is the motivation for approaching the prop, using it, and then discarding it? Early in the rehearsal period you should use the actual prop or a similar substitute so as to become familiar and comfortable in handling it.

In enacting period customs ask yourself similar questions. What is the social significance of the custom? What motivates the character to express himself in this way? How would the character execute the particular custom? For example, in Elizabethan times it was customary for men to remove their hats as a sign of respect to their superiors. A group sent by the Queen to Lord Essex, who was in rebellion against the Queen, removed their hats in the presence of their superior, Essex. Then Essex removed his hat in respect to the Queen's representatives. However, when asked to lay down his weapons he put his hat back on his head as a gesture of contempt. Practice such customs and rituals unitl you can perform them naturally, comfortably, and in character.

MASCULINE STANCE

In general men should avoid standing with their feet together on stage, as this position is static and uninteresting visually. Generally the feet should be spread apart in a forward-backward position or to the side, with the knees straight or with one slightly bent. For the Elizabethan or Restoration period, several stylized positions can be used. The legs can be turned out, with the front heel at the instep of the back foot (the third position). Either the front or the back leg can be bent. For variation, the foot of the bent leg can be lifted so that the ball of the foot rests on the ground. In another stylized period stance (fourth position), both legs are turned out, with the back leg straight and the front leg bent and opened slightly to the side and the heel lifted off the floor.

FEMININE STANCE

The stance of women varies greatly from period to period, depending on fashion and social attitudes. In the theatre, women of the upper classes generally move, stand, and sit with erect posture. Hands and arms are used to make graceful, controlled gestures. In repose the arms and hands are arranged in a decorative way to accent the line of the costume and express the refined sensibilities of the character. Legs rarely are crossed even at the ankle. The general stance and movement patterns suggest containment and control.

Lower-class women by contrast are often characterized by free movement and open stance. Their costumes in period plays are usually less restrictive, since lower-class women must often work. In sitting the legs may be opened to the side or crossed.

DANCE

A choreographer or director will generally guide you through dance requirements in a production. However, a number of techniques will help you to learn the dances faster. Always know which foot you start out on. Note any repeated pattern in the dance. Verbalize the dance movement by movement with words that help you remember the isolated sequences such as "step, kick, bend, rise" or "step, kick, bend, turn." As you move through the actions, say each word in the rhythm required for each action. Once you know the order of the steps, go through them in your mind, and then go physically through the sequence of the steps using very little energy. Then try to go through them up to tempo, and finally try to go through them with the appropriate spirit of the dance. If you are uneasy about dancing at first, remember that although at certain periods a well-bred person was expected to dance well, the period dances were not done by professional dancers as such but by ordinary human beings. It may be helpful to think of learning the dance as a function of the character who is dancing. Ideally a dance furthers the plot, establishes a mood, defines characterization, or establishes the period. It is wise to determine the purpose of the dance and then the style, spirit, and psychology of the period. Consider also the costume being worn, the movements it will dictate, and the way the character would dance, given his social skills, status, and age. In general the dances of the aristocracy in period plays are very refined and dignified, while the peasant dances are spirited and free. This is not always the case, however: for example, Elizabethan court dances, in spite of restrictive clothing, were often very lively affairs, with the queen herself an active participant.

When you are required to dance with a partner, look at pictures to determine how the costume is held when trains are present, what the distance between the partners is, and how the partner is held. Check to see if the partners are facing one another or standing side by side. Study the posture for its distinctive line. Study the relationship of the hands: are they held firmly, or do the woman's fingertips rest on the man's hand? Study the angle of the hand and the elbow, and the carriage of the head. Are props such as handkerchiefs, fans, dance programs, or hats present? If so, how are they held or worn? If you are doing research on a specific dance, determine the customs that preceded or followed it, such as removing the hat, bowing or curtseying, or kissing the lady or her hand.

He who would understand theatre . . . cannot stand only on the stage. He must have one foot in the theatre, one outside.

MORDECAI GORELIK

8 Working on Stage and Camera

In the professional theatre your ability to relate quickly to new ideas and staging may over a period of time mean the difference between being hired or not being hired for a part. There is rarely enough rehearsal time in the theatre, so that any way you can help the production move forward quickly will make you that much more valuable. As so much of the theatre is visual design, including movement, you owe it to yourself to become acquainted with some of the visual requisites of stagecraft. Some of the things that should be considered are the shape of the area you will work in, light and color balance, shapes and forms both static and moving, and your relationship to the overall stage design.

In staging a production, directors may work with one of two approaches, or, more commonly, a combination of them. Some directors demand that every movement and gesture correspond to their predetermined plans, while others encourage free expression and only later work toward setting the staging. For example, one director may tell you exactly when, why, how, and where to move, while another may prefer you to improvise, letting the action evolve naturally and setting it only after a natural sequence has developed. Other directors use a combination of techniques as need or inspiration dictates. All approaches are valid and can produce exciting movement for the stage or camera.

If you are working with an improvisational director, your sense of stage design can help you make valuable contributions to the final staging. Most such directors will be very appreciative of your contributions. If you are working with a director who knows exactly what he wants and directs every detail of the staging, your sense of stage design will enable you to grasp more quickly his intentions and his reasons for staging it in a particular way. Ideally you should be so well prepared that you can respond to the requests of the director whatever his approach.

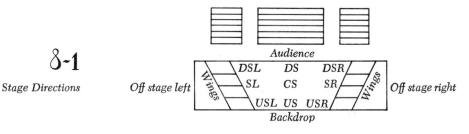

8-1

Stage Directions   *Off stage left*   *Off stage right*

STAGE DIRECTIONS (Figure 8-1)

You should know the common stage terms and be able to respond immediately when given verbal stage directions. Toward the audience is downstage (D.S.). Away from the audience is upstage (U.S.). If you are facing downstage, stage right (S.R.) is to your right and stage left (S.L.) is to your left. Four other common stage directions are downstage right (D.S.R.); downstage left (D.S.L.); upstage right (U.S.R.), and upstage left (U.S.L.). Center stage (C.S.) is in the center of the stage.

An easy way to remember the meaning of D.S. and U.S. is to recall that stages were (and sometimes still are) slanted or "raked" down toward the audience. Hence, moving downstage was literally moving down toward the audience. If you are required to work on a raked stage you will have to adjust your movements accordingly. What seems natural on a flat surface may require new coordination on a slanted surface. Scenery and furniture may also require special design in order to look natural. For example, the front legs of a chair may have to be somewhat longer than those in back.

167

WING NUMBERS

When you are asked to enter on stage or exit off stage there are often specific wing designations. The wings are often set pieces at the sides of the stage with openings between them. They are numbered, with number 1 farthest downstage. The number of wing openings depends in part on the size and kind of stage.

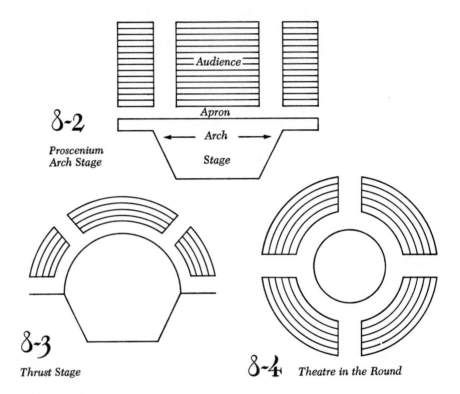

8-2

Proscenium Arch Stage

8-3

Thrust Stage

8-4 **Theatre in the Round**

BASIC STAGE DESIGNS (*Figures 8-2, 8-3, 8-4*)

Stages are designed in many different ways. Some are unalterable. Others are designed so that the playing area and audience can be shifted according to the needs of a particular production. In the permanent theatres there are three basic stage designs. They are the proscenium arch stage, the thrust, open or long-apron stage, and the theatre in the round. Some stages are modified to include elements of all of these designs.

The proscenium arch stage is somewhat like a box with an opening on one side through which the stage action can be viewed. The arch is the frame around the stage picture. Some proscenium arch theatres have an "apron," or stage area just beyond the arch, extending toward the audience. The proscenium arch theatre is generally the easiest stage on which to make entrances and exits. In some proscenium theatres, however, there are stage areas in which the performer cannot be seen by all of the audience. On such stages the performer must adjust his movements to accommodate the "sight lines."

The thrust stage, or a stage with a long apron, generally presents no sight line problems for the actor. However, since exits and entrances are likely to be placed far upstage, actors should time them carefully so as to be in position on cue.

The same problem exists in the theatre in the round, where the performer must often make a long entrance down an aisle. The performer in the round must also learn to play to various sides of the auditorium at different times so that all areas of the audience see him from various angles. There are fewer visual cues for the actor, since scenery is likely to be limited by sight requirements. The theatre in the round requires a special set of directional stage terms. One common method is to designate some area of the stage as twelve o'clock. When you are facing toward the designated area, point one o'clock or two o'clock would be comparable to D.S.R., three o'clock would be comparable to S.R., and so on around the circle. The director can then tell the performer to move, face, exit, or enter from some agreed-on area such as U.S., six o'clock, or aisle six.

If you are touring with a show that moves from theatre to theatre, the director will usually have a "walk through" or staging rehearsal that gives the performers a chance to adjust to the new playing area. Sometimes time schedules and economics do not allow this, in which case you should check the facilities on your own and anticipate any problems that the new area may present. For example, check the size and shape of the stage, the number of wings, the size of the backstage area, whether you can cross from one side of the stage to the other out of view of the audience, and what doors you can use to enter into the backstage area.

WORKING ON CAMERA

Working in live or taped television presents different problems from those on a stage. Working in the cinema is similar to television except that more time can be taken. Television staging is more flexible than theatre staging. There may be more than one camera, and the cameras move as well as the performer, so that he may have to act facing in many directions. The camera with the lighted small red light near the camera lens is the working camera. If you can see the lens you are being "seen" by the camera.

169

The usage of terms is not as standardized as in the theatre. When the performer is facing the camera, the area to the performer's left is "camera right" or "director's right," and the area to the performer's right is "camera left" or "director's left"; the opposite of "stage right" and "stage left." The equivalents of "downstage' and "upstage" are "to the camera" and "away from the camera." "In frame" or "on camera" means "in the picture framed by the camera." "Out of frame" or "off camera" means "outside the picture framed by the camera." Terms vary from studio to studio and person to person.

Entrances and exits for television are likewise considerably more flexible than on the stage and can be made from many different areas. Entrances and exits can be made, for example, from the sides of the playing area or from the camera area. Sometimes it is necessary to "duck under" the camera lens or cross behind the camera to make an entrance from the other side. At other times you may be in place and the camera will move to you. The camera can also move away from you, fade out on you, or cut you off as another camera starts shooting somewhere else. Generally there are monitors around the studio showing the scene currently being taped so that you can check, if necessary, to see that the camera is no longer on you. You may be given a signal by someone in charge as to when you can break.

When you are waiting to move into the playing area, stay clear of other shooting areas, cameras that may have to move quickly, and all the camera cables lying on the floor.

Performing for the camera is generally more intimate than performing on stage. Everything is magnified, so that scenes are played less broadly. While large theatres may require broad facial gestures, television enables even the slightest facial gestures to be seen. Thus in a crowd scene the gestures of supporting players can easily distract attention from the main action if they are not acted appropriately.

Become familiar with the areas to be used in actual performance and learn to gauge your movement patterns so as to move comfortably around the sets. When you are working on stage or on camera, the floor may be taped or painted to show the areas where you are to move and where scenery is to be placed. Be sure you understand these markings thoroughly. Although more freedom is possible on stage than in a television or movie studio, learning to consistently "hit the mark" or repeat exactly the staging that has been set is an important part of the performer's education. On stage it may make the difference between being in the light or in deep shadow. At times even a slight deviation in movement may ruin the dramatic design of a scene.

VISUALIZING STAGE DESIGN

To be more efficient in rehearsal, try to understand some of the visual problems your director must work with and what effects he is after. The director's task is complex. He must coordinate movement of scenery, lighting, costumes, and vocal delivery, and also unify separate scenes so that the total production is dramatically and aesthetically satisfying. Try to sense his aims, so as to help make the production a unified and exciting theatrical event on every level.

Staging a production requires of the director the same sense of design as that of a painter or sculptor. He must sense the best angle from which to view a person or a group. He must decide when reclining, kneeling, sitting, or standing will achieve the most effective stage picture. He must be aware of the effect of placing different bodies next to one another. For example, placing a tall, thin woman next to a short, fat man may have disastrously comic results in a serious drama. He must take into consideration the costumes that will be worn and adjust the spacing and time alotted for movement accordingly. He must also be aware of three-dimensional stage balance and decide which groupings of people will most effectively express the dramatic situation and enhance the visual design of the stage.

At times the director may concentrate on achieving an attractive visual design, or he may group performers to focus attention on important dialogue or action (figure 8-5). Basically there are two kinds of stage balance: symmetrical and asymmetrical (figure 8-6). If the balance is symmetrical both halves of the stage will look the same. A sense of stability will be suggested. Such balance, however, quickly becomes

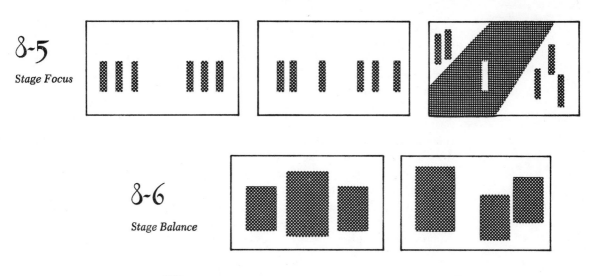

8-5

Stage Focus

8-6

Stage Balance

uninteresting. By proper arrangement of the groups, however, the stage can be harmoniously balanced asymmetrically. In an example of asymmetrical balance, six people in a tight formation in the upper left hand corner of the stage and facing down right might be counterbalanced by a single person in the lower right hand corner facing up left. Such a stage picture is dynamic. What is the relationship between the group and the individual? Does the group pose a threat to the individual, or does the individual control the group? The problem for the director is always to strike a balance between the aesthetic design and the mood of the dramatic conflict in process. Through the judicious use of stage balance, the director can express nonverbally many emotional overtones that words alone cannot convey as effectively.

The director must do more than create static designs, however. The designs must change and develop with the drama. The director must constantly decide if aesthetic stage balance or logical dramatic motivation requires movement by an individual or group. He must also at times decide whether naturalistic or formal movement is more effective. For example, it could be dramatically effective for a group in unison suddenly to inhale and take one step forward while lifting an arm as if to strike out at someone. Unified movement can suggest such disparate ideas as the precision of an army marching against its enemy, the chaos of a lynch mob unified in its anger, or the psychological or thematic overtones of a ritual or ceremony. Such decisions regarding movement are of course related to a director's concept of the style in which the play as a whole is to be presented.

In staging movement, the contrast between static and moving design becomes very important. If the scenery is stationary and on the stage level the director has only to consider the performer's movement design as it relates to a static background. If the set has platforms, elevations, and stairs, coordination of the movement onto or off of the stage and on the various elevations may become more complicated. If the set and performers are both moving, the coordination problem is even more difficult. The problem is compounded because the director generally knows what the set will look like only from blueprints and verbal descriptions. He probably will know if his concepts actually work with the set only a few days before the final dress rehearsal. If the staging does not work with the actual set, the staging may have to be changed. Thus you should be prepared if necessary to learn new movement quickly and completely forget the old staging.

You should be aware of the "strong playing areas" on the stage. These can be altered by selective staging, by set design, and lighting. Not everyone agrees in the assessment of various stage areas, so you should for your own information test the various playing areas for rela tive strength and weakness through observation and analysis.

On a stage without sets or special lighting, direct center is considered by some as the strongest playing area. It is the area of perfect balance and suggests stability. The area is usually reserved for special occasions. Downstage action usually involves the audience more intimately, while scenes played upstage are more removed and thus may be viewed with a more reflective or detached attitude. The downstage left and downstage right areas are considered strong playing areas, as they create an intimate and personal aura. Upstage left and upstage right are also strong playing areas, but their distance from the audience can give a feeling of the nonpersonal or larger-than-life.

The important thing is to realize the particular qualities which certain stage areas may lend to your scenes. Although it is usually the director's prerogative, if you want to build a scene in terms of audience involvement you may develop the movement so that it moves downstage at the climax. Conversely if the scene requires reflective attention or withdrawal, you might move the action upstage. Such decisions also depend on the strengths of the actors involved. Creative directors rarely feel obligated to follow any rigid rules of staging. In general it is wise to involve the audience by continually shifting their visual viewpoint, within the context of the scene. All areas and levels of the stage should be explored and used, just as various postures and bodily movements should be explored and used.

Stage lighting can create subtle shifts in mood, can alter the general shape of the stage by focusing lighting in key areas, and can focus audience attention on the important action in any area. It is important to know what is intended in the lighting of a scene and how you relate to it. Are you to stand in the light or are you to be a silhouette in the shadows of the stage? Does the lighting suggest a mood or state of mind related to the character you are portraying?

In lighting and costuming, color can be used to convey certain ideas or emotions. Color can convey specific moods, aspects of character, states of mind, and visual effects. In addition, colors are sometimes associated symbolically with emotions, moral states, or ceremonies. For example, red is associated with passion, white with purity, purple with royalty, and black with mourning.

Dramatic costuming effects are often achieved through color. Leading performers can stand out even more than usual if dressed in vivid colors or in colors that contrast with those worn by the other performers. Special dramatic effects can be achieved by costuming the entire cast in the same color and painting the backdrop in another color. Costuming

173

different groups of people in different basic colors can help distinguish them by allegiance, symbolic idea, or social class. The director must also consider color balance, which is why he may occasionally place you in what seems an illogical position on stage. Like a painter, he will want to blend or contrast colors for visual effect.

VISUALIZING YOUR OWN BODY ON STAGE

You can be exceedingly helpful to a director in his creation of the overall design of the production. Although it may seem obvious, one of the most important things you can do is to give the director your undivided attention. Listen carefully and avoid unnecessary questions. If you are not rehearsing, either watch the rehearsal to gain a clearer idea of the overall production or rehearse your own part quietly. However, the more you observe how a production is put together, the quicker a study you can become. The more good or bad performing you can observe, the more potential you have for improving your own performance.

As you are being blocked into a scene, try to understand how you fit into the overall stage design. Determine how your body position relates to those of the other performers, the furniture, and the sets. For example, if the set is constructed with a vertical thrust upward, you may want to work posturally to emphasize or to contrast with it. You may wish to use doorways or rectangular shapes to frame yourself interestingly. Be aware of problems of proportion between yourself and a realistically painted backdrop. Use furniture to visual advantage. If you are near a table, for example, consider how you can position your body to create the most exciting and relevant visual design. Behind it? Leaning over it? Using it as a barrier between you and another character? If you are working on stairs, ramps, or platforms, determine how you can position your body to enhance the shape of the elevation or vary your space relationships with other performers. Such problems can be worked out in cooperation with directors.

It is equally important to think of yourself in physical relation to others on the stage. A straight line of actors facing the audience is very uninteresting. The dynamics of a scene must be reflected in dynamic movement which illustrates relationships between the characters. Thus the focal character may be upstage or, conversely, may be downstage facing the audience. At moments of crisis certain characters may adopt body positions (crossed arms, for example) for which others will have to compensate.

Speaking and moving effectively as a crowd requires detailed direc-

174

tion but also great ensemble sensitivity from each of the actors involved. At the same time that you are carefully listening to individual actors, you must as a group modulate and time your responses as the scene requires. In the same way, the physical responses of a crowd must be orchestrated and modulated.

In a naturalistic mob scene, individuals united in a common mood will nevertheless usually exhibit individual movement patterns and postures. However, the dramatic or symbolic intention of a scene may require the mob to face or gesture in the same way so as to suggest some mutually felt emotion. If that is so, try to move exactly in unison with the other actors. A single error in timing or action can destroy the illusion of group action.

At other times a crowd may be asked to "dress the stage" or to "mill." To dress the stage is to make sure that there are not large empty areas on it, and to mill is to move around naturally and improvisationally in the context of a scene. If the scene requires unstructured milling about, then singly or in couples the crowd can move from place to place speaking to people or groups as they pass. The important thing is to sense what part of the stage is inactive, to move naturally from place to place, and to give a sense of reality to what you are doing.

Some performers have difficulty staying in character when they are not either speaking or moving. As a result they fail to give proper support to the other performers. There are several things that can be done to prevent this happening. As an actor, you must listen and watch as intently as you would in real life — and sometimes moreso to maintain the focus of the scene. This will show the audience where the main action is taking place. Listen as if you had never heard the scene before, and occasionally shift your stance, your head or your arms, as you might if resting in real life. This can prevent the body from looking wooden or frozen. However, it should never be done in such a way that it distracts from the main action. Another method is to lift the heels imperceptibly off the floor; you will automatically stay alert to keep your balance.

Audibility can be a major problem for an actor in an acoustically poor room or outdoors. The first defense is to speak more loudly and distinctly. However, the actor can also provide helpful visual cues. In a crowd, for example, a new speaker can draw attention to himself by gesturing or moving slightly forward. If the dialogue is passed back and forth between several characters it is helpful to focus attention by placing them somewhat forward of the group. In general when you are speaking or singing your head should be directed more or less toward the audience. As a rule, avoid speaking into the floor, directly into the wings, or upstage. If required to do so, be sure to speak distinctly.

One method for developing the sense of spatial balance is carefully to observe people, nature, photographs, paintings, or the arrangement of inanimate objects like furniture in a room, and then to analyze what makes them visually interesting or uninteresting as groupings. Ask such questions as: What emotions do the groupings suggest? What elements of line, size, shape, weight, color, and light are present, both in the objects themselves and in the empty spaces that surround them? To what focal point is your eye drawn? How might some of these arrangements be effectively utilized in a theatrical setting?

Other useful methods are used by art students. Gather from around the house a variety of small objects that vary in size, shape, weight texture, and color, and then arrange them in different patterns that explore the possibilities of symmetry, asymmetry, and spatial depth. Using a flashlight or table lamp, light the objects from behind, from in front, from the sides, from overhead, or if possible from below. In each case study the shifting effects of the light on the balance of the objects.

Studying magazine and newspaper pictures can also be helpful. They often sum up the dramatic essence of some situation. Select a picture with several people in it. Using colored crayon, draw a line in one color through the main flowing action of a single figure. With another color draw a line to join the group together as a spatial unit, and with another relate them to their setting. Studying the lines and shapes that unite the picture can give you some feeling for stage design.

Look at television, movies, and stage plays critically to see how others have used the individual and groups in poses and in moving. From what directions was a movie or television scene shot and why? Why were fadeouts, special angles, close-ups or pans (moving the camera closer or farther away from the performer) used? Why was there a close-up shot of someone handling a prop?

In looking at a stage production, imagine that you are taking a series of still photos. At any given moment what is the overall stage picture? How has the stage been balanced with performers, set, lights, color, and costumes? What shapes are created by the "empty spaces" on stage: that is, the areas between the performers or between set pieces? How skillfully do the performers move? What are the strong or weak playing areas and why?

Other activities that can be done at a party or alone will help to develop your awareness of the effects of space on yourself and others. For example, in various rooms analyze your reaction to sitting in a corner, lying under a table, standing in the center of the room, or elevating yourself so that you can look down on the room. Try to see each room from as many new perspectives as possible.

Visit rooms crowded with people or furniture and rooms that are completely empty. Visit buildings where the rooms are very small with low ceilings or very large with high ceilings. Walk on very crowded streets, along deserted open spaces such as beaches, and in enclosed spaces such as forests. In each case try to sense and compare your feelings. Try to determine why you react to the space the way you do.

Observe how people use and react to different areas of space. What do their reactions reveal about them? What part of the room does a given person choose to occupy? Does he sit or stand in the center of the room, for example, or in a corner of the room, or with his back to the wall? If he is given the choice, does he generally select the same part of the room each time he returns to the room? How does he relate to the room and others in it? Does he move freely and with self-assurance, relax comfortably back into a chair, or sit erectly on the edge of the furniture? Does this have anything to do with the size of the room or the shape of the furniture?

If possible, observe the same person in different rooms in different situations. How does his movement and use of space change, for example, in a kitchen, a living room, an office, a large or small room, or a church?

People subjected for long periods of time to crowded situations eventually react to the stress of the situation with increased irritability. Try to observe people in crowded situations of different kinds and determine which tend to produce the most irritability. Note how people react to the situation with facial gestures, body gestures, and actions. For example, how do they act in a crowded supermarket at the checkout lines, on a crowded bus, on a busy freeway or expressway during rush-hour traffic, at a crowded theatre when the event has just ended, or at a crowded beach on a pleasant afternoon?

Afterword

It has not been my intention to write a rule book for stage movement that outlines exact details of how, when, and why to move, but rather to suggest some approaches to movement and movement analysis in the hope that they will provide a basis for you to explore, observe, and develop your own artistic viewpoint.

The artist is more than a human camera that simply records what is perceived. The artist in the practice of his craft intuits something, distills it, refines it, and polishes it so that he is able through his voice and body to speak to his audience, whether they are children or adults, sophisticated or unsophisticated, learned or unlearned.

The artist does not settle for the tried and true formula that may guarantee some kind of success. The artist is always vitally alive, aware, and open to new ideas, new techniques, and the perfection of his craft. The artist taps the resources of his unconsciousness — feelings, impressions, intuitions, and perception — and tries to speak with his whole being. Although the path traveled by artists is sometimes long and arduous, the journey is filled with great rewards.

Selected References

MOVEMENT ANALYSIS

    *Alexander, F. Matthias. *The Resurrection of the Body,* ed. Edward
        Maisel. New York: Dell, 1971.

    Davis, Martha. *Understanding Body Movement: An Annotated Bibli-
        ography.* New York: Arno Press, 1972.

    *Dell, Cecily. *A primer for Movement Description: Using Effort-Shape
        and Supplementary Concepts.* New York: Dance Notation Bureau,
        1970.

    Feldenkrais, Moshe. *Awareness Through Movement: Health Exercises
        for Personal Growth.* New York: Harper and Row, 1972.

*Signifies paperback edition.

181

*Humphrey, Doris. *The Art of Making Dances,* ed. Barbara Pollack. New York: Grove Press, 1959.

*Jaques-Dalcroze, Emile. *Rhythm, Music and Education,* trans. Harold F. Rubinstein. New York: Benjamin Blom, 1972.

Laban, Rudolf. *Choreutics,* ed. Lisa Ullman. London: MacDonald and Evans, 1966.

——————. *Modern Educational Dance.* New York: Frederick A. Praeger, 1968.

—————— and Lawrence, F. C. *Effort.* London: MacDonald and Evans, 1947.

Logan, Gene A. and Mckinney, Wayne C. *Kineseology.* Dubuque, Iowa: Wm. C. Brown, 1970.

Royal Canadian Air Force Exercise Plans for Physical Fitness. Rev. Ed. United States: Pocket Books, 1962.

*Shawn, Ted. *Every Little Movement: A Book About Francois Delsarte.* 2nd ed.; Brooklyn: Dance Horizons, 1963.

*Todd, Mabel Elsworth. *The Thinking Body: A Study of the Balancing Forces of Dynamic Man.* Brooklyn: Dance Horizons, 1937.

PERFORMANCE, PREPARATION AND THEORY

*Artaud, Antonin. *The Theatre and its Double.* trans. Mary Caroline Richards. New York: Grove Press, 1958.

Benedetti, Robert L. *The Actor at Work.* Englewood Cliffs, New Jersey: Prentice-Hall, 1970.

Brook, Peter. *The Empty Space.* New York: Atheneum, 1968.

Chekhov, Michael. *To the Actor: on the Technique of Acting.* New York: Harper and Row, 1953.

Chisman, Isabel and Raven-Hart, Hester Emilie. *Manners and Movements in Costume Plays.* London: H. F. W. and Sons, n. d.

Dannett, Sylvia G. L. and Rachel, Frank R. *Down Memory Lane: Arthur Murray's Picture Story of Social Dancing.* New York: Greenberg. 1954.

Enters, Angna. *On Mime.* Middleton, Connecticut: Wesleyan University Press, 1965.

*Grotowski, Jerzy. *Towards a Poor Theatre.* New York: Simon and Shuster, 1968.

*Hethmon, Robert H., ed., *Strasberg at the Actors Studio: Tape Recorded Sessions.* New York: Viking Press, 1965.

Hobbs, William. *Stage Fights, Swords, Firearms, Fisticuffs, and Slapstick.* New York: Theatre Arts Books, 1967.

182

Hodgson, John and Richards, Ernest. *Improvisation: Discovery and Creativity in Drama.* London: Methuen and Co., 1966.

*King, Nancy. *Theatre Movement: The Actor and his Space.* New York: Drama Book Specialists, 1971.

Laban, Rudolf. *The Mastery of Movement,* rev. Lisa Ullman, 3rd ed.; London: MacDonald and Evans, 1971.

Mawers, Irene. The Art of Mime, 5th ed.; London: Methuen, 1960.

*Moore, Sonia. *The Stanislavski System: The Professional Training of an Actor.* New York: Pocket Books, 1967.

Oxenford, Lyn. *Design for Movement: A Textbook on Stage Movement.* New York: Theatre Arts Books, 1951.

Shepard, Richmond. *Mime: The Technique of Silence.* New York: Drama Book Specialists, 1971.

Stanislavski, Constantin. *Building a Character,* trans. Elizabeth Reynolds Hapgood. New York: Theatre Arts Books, 1949.

*White, Edwin and Battye, Marguerite. *Acting and Stage Movement.* New York: Arc Books, 1963.

*Whitman, Robert F. *The Play Readers Handbook.* New York: Bobbs-Merrill, 1966.

BEHAVIOR AND EXPRESSION

*Berne, Eric. *What Do You Say After You Say Hello: The Psychology of Human Destiny.* New York: Bantam, 1973.

*Birdwhistell, Ray L. *Kinesics and Context; Essays on Body Motion Communication.* New York: Ballantine, 1972.

Darwin, Charles. *The Expression of the Emotions in Man and Animals.* London: Murray, 1872.

Eibl-Eibesfeldt, Irenaus. *Love and Hate: On the Natural History of Basic Behaviour Patterns,* trans. Geoffrey Strachan. London: Methuen and Co. 1971.

*Fast, Julius. *Body Language.* New York: Pocket Books, 1971.

*Feldenkrais, M. *Body and Mature Behavior: A Study of Anxiety, Sex, Gravitation and Learning.* New York: International Universities Press, Inc., 1949.

*Hall, Edward T. *The Silent Language.* New York: Fawcett, 1966.

*Knapp, Mark L. *Nonverbal Communication in Human Interaction.* New York: Holt, Rinehart and Winston, Inc., 1972.

Lange, Carl G. and James, William. *The Emotions.* New York: Hafner (facsimile of 1922 ed.), 1967.

*Lewis, Howard R. and Streitfeld, Harold S. *Growth Games.* New York: Bantam, 1972.

183

*Lorenz, Konrad. *On Aggression,* trans. Marjorie Kerr Wilson. New York: Bantam, 1969.

*Mead, Margaret. *Male and Female: A Study of the Sexes in a Changing World.* New York: Dell, 1968.

*Morris, Desmond. *The Naked Ape.* New York: Dell, 1969.

*Nierenberg, Gerard I. and Calero, Henry H. *How to Read a Person Like a Book.* New York: Pocket Books, 1973.

*Perls, Frederick, Hefferline, Ralph F., and Goodman, Paul. *Gestalt Therapy: Excitement and Growth in the Human Personality.* New York: Dell, 1951.

*Sachs, Curt. *World History of the Dance,* trans. Bessie Schonberg. New York: W. W. Morton, 1963.

Wolff, Charlotte. *A Psychology of Gesture,* trans. Anne Tennant, 2nd ed.; London: Methuen and Co., 1948.

MANNERS, MOVEMENT, AND CLOTHING

Boehn, Max Von. *Modes and Manners,* trans. Joan Joshua, 4 vols. Philadelphia: J. B. Lippincott, 1932, 1933 and n.d.

Boehn, Max Von and Fischel, Oskar. *Modes and Manners of the Nineteenth Century,* trans. M. Edwards, 2 vols. New York: Benjamin Blom, 1970.

Boucher, Francois. *20,000 Years of Fashion: The History of Costume and Personal Adornment.* New York: Harry N. Abrams, n.d.

Broby-Johansen, R. *Body and Clothes,* trans. Erik I. Friis and Karen Rush. New York: Rheinhold, 1968.

*Wood, Melusine. *Historical Dances (Twelfth to Nineteenth Century): Their Manner of Performance and Their Place in the Social Life of the Time.* London: The Imperial Society of Teachers of Dancing, 1964.

Index

Pain, 29, 41
Pantalettes, 159
Pantalone, 137
Pantomime, 35-38
Parkinson's Disease, 66
Passion, 156
Passivity, 16
Patriotic gesture, 62
Pedaling, 85
Pedrolino, 137
Pelvis, 25
Percussive action, 13, 14, 15
Period movement, 152-65
Period plays, 26, 56, 57, 58, 68, 100, 106, 161, 165
Peripheral vision, 75, 101
Personality profile, 137-41
Personality types, 120-23
Phrase expression, 98
Pierrot, 137
Platforms, 172, 174
Playing areas, 172-73
Poise, 55, 72
Political gesture, 62
Posture, 11, 30, 54-56, 58-60, 70, 139-40. *See also* Carriage
Prayer, 62
Pregnancy, 125
Preparation, 134
Pressure, 29, 30
Pride, 25, 44, 56, 70
Primary design, 52
Props, 35-38, 144-45, 162-63
Proscenium arch stage, 168
Psychology of Gesture, A, 63
Puberty, 124
Pulling, 13, 14, 15
Punching actions, 110
Purity, 156
Pushing, 13, 14

Rage, 42. *See also* Anger

Rapier, 109, 163
Reflection, 44, 72
Rehearsal, 163, 166, 169
Relaxation, 12-16
Religious gesture, 62
Renaissance, 26
Resistance, 13-16 *passim*
Restoration, 137, 161-62, 163-64
Reverence, 44
Rhythm, 12, 16-21, 69, 107-8, 109, 118, 149
Royalty, 70
Rural patterns of movement, 131

Scene-stealing, 151
Secondary design, 52
Self-awareness, 89
Sensation attributes, 30
Senses, 2-3, 29-35, 88-94. *See also* Hearing; Kinesthetic sense; Sight; Smell; Static sense; Taste; Touch; Visceral sense
Sensitivity training, 31-35, 89
Sensory awareness, 30
Sets, 174
Sexual differences, 127-28
Shaking, 7, 13, 15, 78. *See also* Vibratory action
Shame, 44, 49
Shape exercises, 97-98
Shoes, 64, 158
Shoulders, 25, 84
Shyness, 25, 44, 49, 57, 70
Sight, 29, 30, 33-34
Silhouette, 24-25, 55, 156-57
Singing, 150-51
Sissonne, 106
Sitting, 58-60, 125, 143-44, 160, 163

Sitting exercises, 86-88
Skeleton, 1-2
Skin, 2, 126
Skin sensations, 29
Skirt, 160
Skirt, hooped, 159-60
Slash action, 13, 15, 110
Sleeves, 158
Sliding, 13, 15
Slyness, 44
Smell, 29, 30, 34, 44, 63, 88-89
Snuff boxes, 162-63
Sobbing, 40. *See also* Crying
Social attitudes, 140
Sorrow, 40, 41, 126
Space, 21-24, 118
Spatial balance, 176-77
Spatial design, 22-24
Speaking, 175
Speed, 16, 68-69
Spine, 125
Spontaneity, 133
Sports shows, 21
Stage design, 171-74, 176
Stage directions, 72, 167, 169
Stage fights, 109-13
Stagecraft, 166
Stairs, 72, 172, 174
Stance, 55, 125, 163-64
Standing, 144
Standing exercises, 78-84
Static sense, 3, 29, 30
Step-sways, 83
Stick fighting, 111-12
Stock characters, 137
Stretching, 83-84
Stroke, 66
Stumble falls, 114
Supporting exercises, 90-91
Surprise, 48
Suspension, 13-14, 15
Suspicion, 25, 42

188

C90026380/

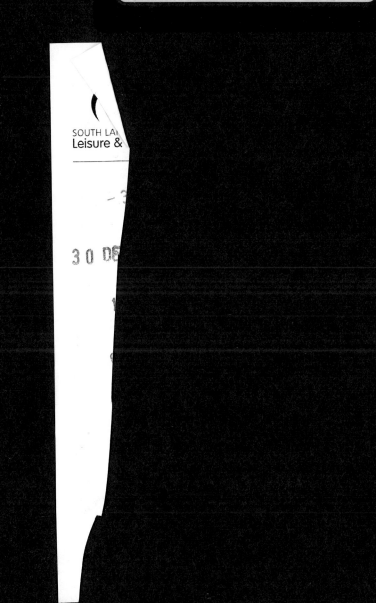

SOUTH LA
Leisure &

3 0 DE